Lecture Notes in Chemistry

Edited by G. Berthier, M. J. S. Dewar, H. Fischer
K. Fukui, G. G. Hall, H. Hartmann, H. H. Jaffé, J. Jortner
W. Kutzelnigg, K. Ruedenberg, E. Scrocco, W. Zeil

21

Kurt Varmuza

Pattern Recognition in Chemistry

Springer-Verlag
Berlin Heidelberg New York 1980

Author

Kurt Varmuza
Institut für Allgemeine Chemie
der Technischen Universität Wien
Lehargasse 4
A-1060 Wien

ISBN-13:978-3-540-10273-1   e-ISBN-13:978-3-642-93155-0
DOI: 10.1007/978-3-642-93155-0

© by Springer-Verlag Berlin Heidelberg 1980

2152/3140-543210

Preface

Analytical chemistry of the recent years is strongly influenced by automation. Data acquisition from analytical instruments – and sometimes also controlling of instruments – by a computer are principally solved since many years. Availability of microcomputers made these tasks also feasible from the economic point of view. Besides these basic applications of computers in chemical measurements scientists developed computer programs for solving more sophisticated problems for which some kind of "intelligence" is usually supposed to be necessary. Harmless numerical experiments on this topic led to passionate discussions about the theme "which jobs cannot be done by a computer but only by human brain ?". If this question is useful at all it should not be answered a priori. Application of computers in chemistry is a matter of utility, sometimes it is a social problem, but it is never a question of piety for the human brain.

Automated instruments and the necessity to work on complex problems enhanced the development of automatic methods for the reduction and interpretation of large data sets. Numerous methods from mathematics, statistics, information theory, and computer science have been extensively investigated for the elucidation of chemical information; a new discipline "chemometrics" has been established.

Three different approaches have been used for computer-assisted interpretations of chemical data. 1. Heuristic methods try to formulate computer programs working in a similar way as a chemist would solve the problem. 2. Retrieval methods have been successfully used for library search (an unknown spectrum is compared with a spectral library). 3. Pattern recognition methods are especially useful for the classification of objects (substances, materials) into discrete classes on the basis of measured features. A set of characteristic features (e.g. a spectrum) of an object is considered as an abstract pattern that contains information about a not directly measurable property (e.g. molecular structure or biological activity) of the object. Pure pattern recognition methods try to find relationships between the pattern and the "obscure property" without using chemical knowledge or chemical prejudices.

This book gives in the first part an introduction to some pattern recognition methods from a chemist's point of view. This introduction is by no means systematical or exhaustive but has been restricted to simple mathematical methods. No previous knowledge of pattern recognition should be necessary for the reader.

Chapter 1 gives some basic ideas of pattern recognition; this Chapter may be skipped by readers already familiar with pattern recognition.

Chapters 2 to 8 describe several pattern recognition methods with emphasis on binary classifiers and methods already applied in chemistry. Each Chapter is preceded by an introductory part showing the principles of the method. These Chapters should be readable in arbitrary sequence.

Chapters 9 and 10 deal with preprocessing of original data and feature selection. These problems must be treated at the beginning of a pattern recognition application. These Chapters have not been positioned at the beginning of the text because a more detailed description of these subjects requires some basic knowledge of pattern recognition methods.

Chapter 11 deals extensively with objective methods for evaluating pattern classifiers - a subject which has been often neglected in chemical applications of pattern recognition methods.

After reading appropriate Chapters of the first part of this book a reader with some knowledge in computer programming should be able to apply simple versions of pattern recognition methods to actual problems without further studies.

The second part of this book presents a description of reported applications in chemistry. The aim of these Chapters was completeness; however, the boundaries of chemistry are diffuse and the papers are spread over many journals. It was not always possible to judge the actual merit of pattern recognition methods in distinct fields of chemistry; conclusions should be drawn by the reader himself considering specific and varying demands and previous knowledge of actual classification problems.

Chapter 12 gives an overview about pattern recognition applications in chemistry. Chapters 13 to 20 extensively describe applications in spectral analysis, chromatography, electrochemistry, material classification, structure-activity-relationship research, clinical chemistry, environmental chemistry and classification of analytical methods.

A comprehensive list of literature references, an author and subject index should facilitate a retrieval of detailed and original information about pattern recognition in chemistry. All cited literature except a few has been used in original form.

Further developments on this field are necessary. It would be an honor for the author if this book was an aid to or stimulant of the reader.

Vienna, April 1980   K. Varmuza

Technical Remarks

This text has been typed by the author himself using a self-written simple text-editor-program running on a mass spectrometric data system. Therefore the author is responsible for all errors in the manuscript.

For technical reasons <u>vectors</u> are denoted by over-lined characters; e.g. $\bar{x}$ is a vector with components x_1, x_2, ... but not a mean value !

The <u>reference list</u> and author list have been handled by a couple of self-written Fortran programs running at the Computer Centre of the Technical University of Vienna. The author apologizes for some unusual notations.

Acknowledgements

Many thanks to my colleague and friend Dr. Heinz Rotter for stimulating and critical discussions. Also thanks to Harold Urban for the drawings.

C O N T E N T S

Part A

Introduction to Some Pattern Recognition Methods

1. Basic Concepts

1.1. First Ideas of Pattern Recognition

In scientific work the problem of determining or predicting an obscure property of an object or an event often arises. The property is called obscure because only indirect measurements can be made that are known or supposed to be related to the property in question. If a theoretical relationship between the measurements and the property is not completely known a method of data interpretation may be used which is sometimes called "educated guess" [153]. In this approach an empirical relationship is derived from a collection of objects for which the interesting property and the indirect measurements are known. This relationship is then used to predict the obscure property of unknown objects.

A set of indirect measurements which describe one object is called a pattern. The determination of the obscure property is often a recognition of the class (category) to which a pattern belongs. Classification of patterns is a fundamental process in many parts of science and human being and therefore mathematical methods of pattern recognition find wide applications in very different fields.

An example from chemistry may characterize a typical pattern recognition problem: The objects may be chemical compounds and the property to be determined is the presence of a carbonyl group in the molecule. This is an obscure property which can be recognized only by indirect measurements. It is known that the mass spectrum is related to the molecular structure but no complete theoretical understanding of this relationship exists today. A pattern recognition approach to this problem requires a large set of mass spectra (from compounds with a carbonyl group and without a carbonyl group). Features are derived from the spectra that are supposed to be related to the problem. Now an empirical relationship is derived between the features and the presence of a carbonyl group. The relationship is formulated as a classification procedure (a classifier) and is tested on a set of known compounds. If the classifier is able to detect a carbonyl group with a high rate of success it may be also useful for unknown samples (objects). A contrary result of this investigation may be that there is no satisfactory relationship between the chosen mass spectral features and the presence of a carbonyl group in the molecule.

The purpose of pattern recognition is to categorize a sample of observed data as a member of the class to which it belongs. Pattern recognition is a general solving tool to help when direct and theoretical approaches fail. Non-trivial methods are necessary for pattern classification because more than two measurements are usually used to describe an object. But at least in the area of chemistry most of the successful methods of pattern recognition are conceptually simple and the understanding and application of the methods do not require an extensive knowledge of mathematics or statistics.

1.2. Pattern Space

Many pattern recognition methods can be explained by an intuitive geometric description of the classification problem. This point of view will be used throughout this book.

The basic concept is the following: An <u>object</u> or an event j is described by a set of d <u>features</u> x_{ij} (i = 1...d). All features of one object form a <u>pattern</u>. For a simpler description of the method one may assume that only two features (measurements) x_{1j} and x_{2j} are known for each object j. In this case, the object can be represented as a point in a 2-dimensional coordinate system (<u>pattern space</u>). The features correspond to the coordinate axis and the numerical values of the features are used as coordinates. An equivalent representation is a vector $\bar{x}_j$ (<u>pattern vector</u>) from the origin to the point with the coordinates x_{1j} and x_{2j}.

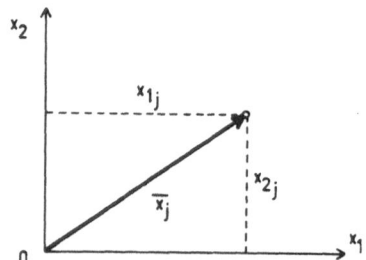

FIGURE 1. An object is characterized by two measurements x_{1j} and x_{2j} and represented in a 2-dimensional pattern space by a point with the coordinates x_{1j}, x_{2j} or by a pattern vector $\bar{x}_j$.

The <u>hypothesis for all pattern recognition methods</u> is that similar objects - similar with regard to a certain property - are close together in the pattern space and form clusters.

A desirable case is shown in Figure 2. All considered objects form two distinct clusters and the membership to a cluster corresponds to a physically meaningful property (+ or -). Classification of an object (o) whose class membership is unknown requires the determination of the cluster to which this point belongs.

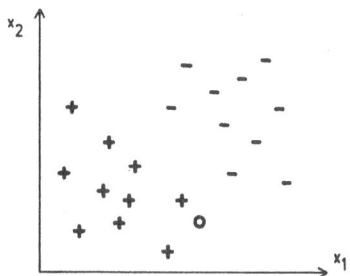

FIGURE 2. Two-dimensional pattern space with two distinct clusters for the classes 'plus' (+) and 'minus' (-). The unknown (o) is classified to one of the two classes.

In practical applications, a pattern space with much more than two dimensions is necessary. Clustering in a d-dimensional <u>hyperspace</u> is of course not directly visible or imaginable to the scientist. In this case, pattern recognition methods are helpful.

Mention of a d-dimensional hyperspace (d greater than 3) gives rise to many chemists to <u>finish</u> their <u>interest</u> in pattern recognition. Therefore, it seems necessary to point out that there is no qualitative difference between the geometry in a d-space and the geometry in 2 or 3 dimensions - the difference in computation is only quantitative. All methods and formulas will be derived from 2-dimensional examples. The extension to more dimensions is merely a formalism and does not require thinking in more than three dimensions.

A very important problem in pattern recognition is the selection of the features which are used to characterize an object or an event. In the first and most important step chemical knowledge must be used to select those measurements which are hopefully related to the classification problem. The measurements for an object may stem from one

type of instrument (<u>single-source data</u>) or from different types of
instruments (<u>multiple-source data</u>). After an heuristic pre-selection
of features several mathematical methods (<u>feature selection</u>) are available to find the best features for a given classification problem.
Preprocessing of the numerical values of the features is used to improve clustering and/or to facilitate the development and application
of classifiers.

1.3. Binary Classifiers

In a binary classification problem one has to distinguish only
between <u>two mutually exclusive classes</u> (e.g. class 1 contains compounds
with a certain chemical substructure, and class 2 contains all other
compounds). If the two classes form well separated clusters in the pattern space it is often possible to find a plane (<u>decision plane</u>) which
separates the classes completely (Figure 3). In this case, the data are
called to be "<u>linearly separable</u>". The calculation of a suitable decision plane is often called the "<u>training</u>".

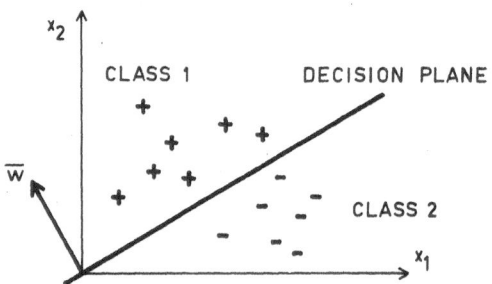

FIGURE 3. A decision plane (a straight line in this 2-dimensional
example) separates the two classes of objects and is defined by a
decision vector $\bar{w}$.

The decision plane is usually defined by a <u>decision vector</u> (<u>weight
vector</u>) $\bar{w}$ orthogonal to the plane and through the origin. The weight
vector is very suitable to decide whether a point lies on the left or

right side of the decision plane (t.m. whether it belongs to class 1 or 2). The sign of the <u>scalar product</u> (<u>dot product</u>) s which is calculated from weight vector $\bar{w}$ and pattern vector $\bar{x}$ is positive for class 1 and negative for class 2 (Figure 4).

$$s = \bar{w}.\bar{x} = |\bar{w}|.|\bar{x}|.\cos \alpha \tag{1}$$

For more than two dimensions another equivalent equation for the scalar product is more convenient:

$$s = \sum_{i=1}^{d} w_i x_i \tag{2}$$

$$s = w_1 x_1 + w_2 x_2 + \ldots + w_d x_d$$

s : scalar product (dot product)
$x_1, x_2, \ldots x_d$: components (measurements, features) of the pattern vector
$w_1, w_2, \ldots w_d$: components of the weight vector (decision vector)
d : number of dimensions

The development of a decision vector is usually computer time consuming. But the application of a given decision vector to a concrete classification requires only some multiplications and summations which can be easily done even by a pocket calculator.

Linear binary classifiers (Figure 5) - defined by a decision vector and the scalar product algorithm - are the most used type of classifiers in chemical applications of pattern recognition methods. Computational methods for binary classifiers are extensively discussed in Chapter 2. A <u>multicategory classification</u> can be performed by a series of successive binary classifications or by special multicategory classifiers (Chapter 2.6.3).

Many considerations in the d-dimensional hyperspace are simplified if the decision plane and the decision vector pass through the origin. Such a decision plane was possible in the special case shown in Figure 1, but an extension of the data is necessary for general cases (Figure 6).

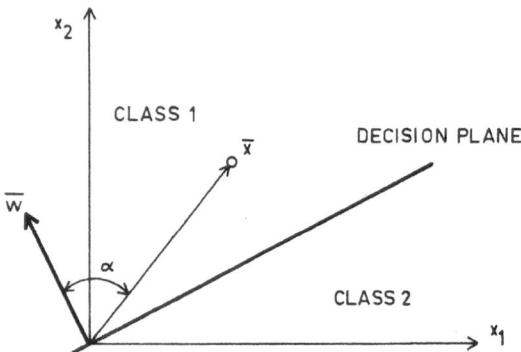

FIGURE 4. The sign of the scalar product from weight vector $\bar{w}$ and pattern vector $\bar{x}$ is positive for class 1 and negative for class 2.

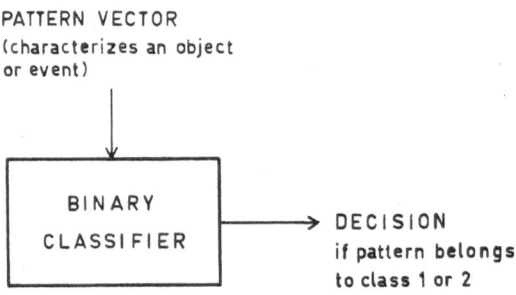

FIGURE 5. Principle of a binary classifier.

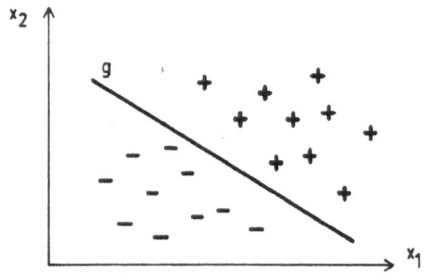

FIGURE 6. No decision plane through the origin is possible.

If all pattern vectors are augmented by an additional component
(with the same arbitrary value in all patterns), then a decision plane
which passes through the origin is possible as shown in Figure 7.

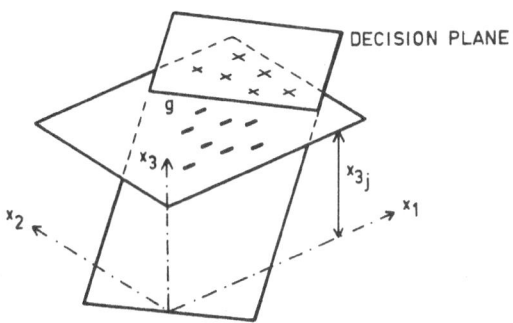

FIGURE 7. Augmentation of all pattern vectors by an additional component
x_{3j} (same value in all patterns) makes possible a decision plane through
the origin.

1.4. Training and Evaluation of Classifiers

Figure 8 shows a widely used procedure for the development and
evaluation of binary classifiers. A collection of patterns with known
class memberships (e.g. a library of spectra) is randomly divided into
two parts. Part 1 is used as a training set to develop a classifier that
recognizes the class membership of the training set patterns as well as
possible. The percentage of correctly classified training set patterns
is called recognition rate [128]. The classifier is then tested with the
patterns of the second part which is called prediction set. The percentage of correctly classified prediction set patterns (which have not been
used during the training) is called predictive ability P. Recognition
rate and predictive ability are only preliminary and sometimes confusing
criteria of the performance of a classifier. An extensive discussion of
evaluation problems is given in Chapter 11.

An alternative procedure for an investigation of pattern recognition
methods is the leave-one-out-method shown in Figure 9. At any given time

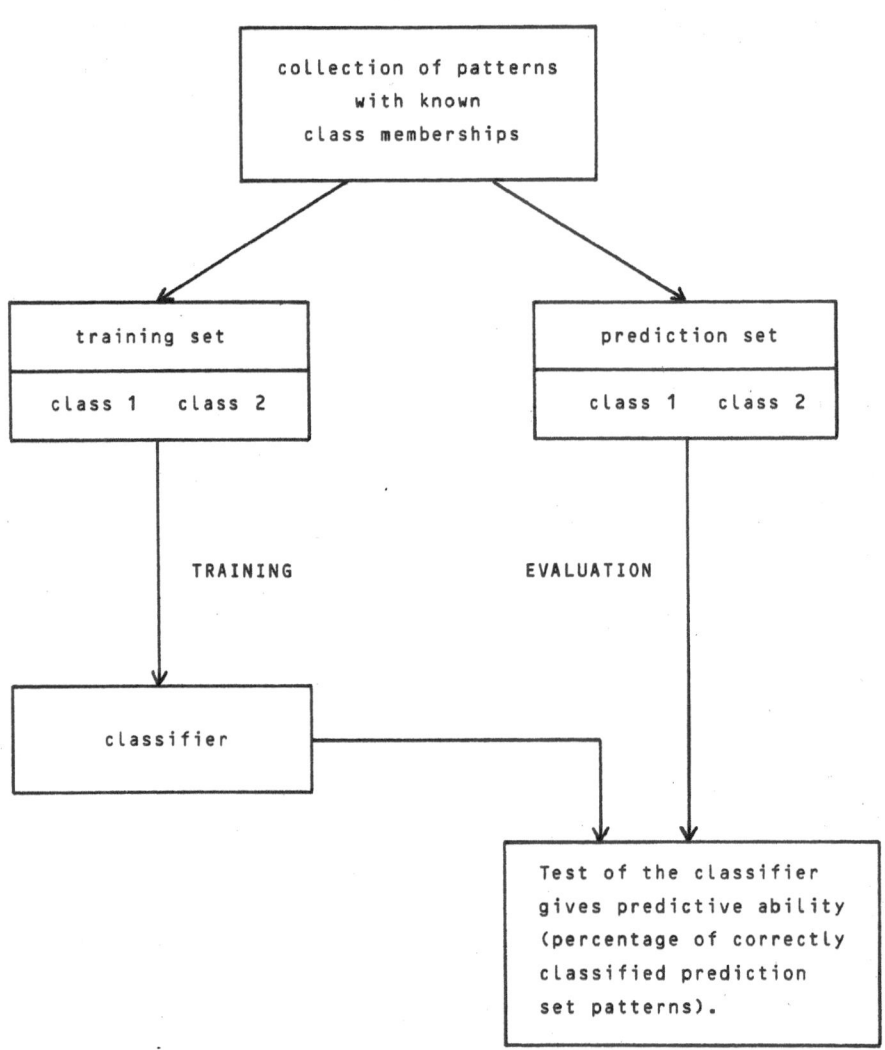

FIGURE 8. Procedure for training and evaluation of a binary classifier.

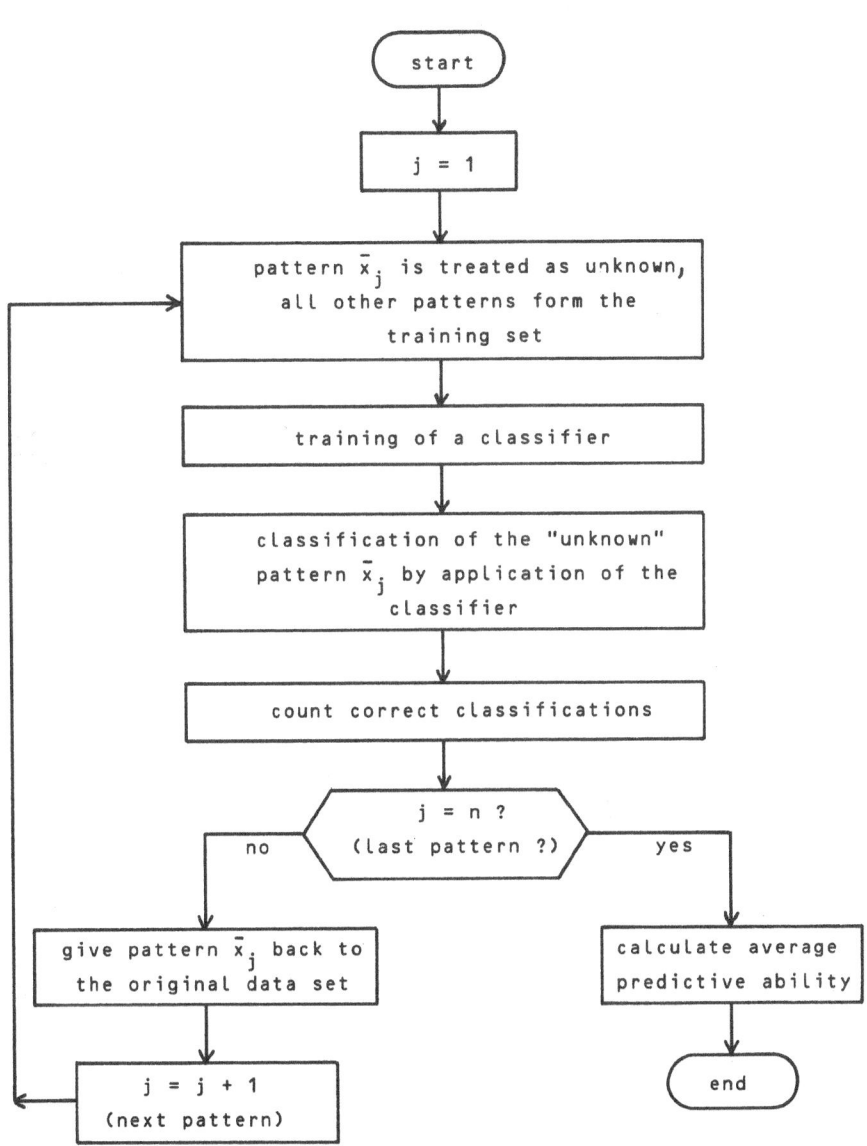

FIGURE 9. Leave-one-out-method for an investigation of pattern recognition methods. n: total number of patterns, j: running variable for
pattern number (j = 1...n).

only one pattern is considered as unknown and all other patterns form
the training set. Training and prediction is repeated until each pattern
was treated as unknown once. From all predictions an averaged predictive
ability is calculated [292]. This procedure is very time-consuming because at each step a new classifier has to be trained. Less laborious
methods are the leave-ten-out-method [289], or the leave-a-quarter-of-
the-patterns-out-method [81].

1.5. Additional Aspects

A part of pattern recognition deals with the recognition and
analysis of clusters in the d-dimensional pattern space. Various methods
are available to find all points belonging to the same cluster (Chapter 7). Display methods trie to project or map the d-dimensional space
onto 2 or 3 dimensions (Chapter 8). The human interpreter with his enor-
mous pattern recognition capability may separate clusters on the projection sometimes better than a computer program.

If the class membership of the patterns are known one speaks about
supervised learning (learning with a teacher). More difficult is the
problem if no natural classes of patterns are known a priori. In this
case an unsupervised learning (learning without a teacher) requires a
cluster analysis and a physically meaningful interpretation of the clusters.

Most pattern recognition methods are nonparametric. That means it
is not necessary to know the underlying statistic (e.g. the probability
for a class at a certain point or region of the pattern space).
Parametric methods (Chapter 5) assume a knowledge of the class-conditional probabilities (which are very hard to estimate in practical problems).

1.6. Warning

The results of many pattern recognition applications in chemistry
are doubtful because an important prerequisite of the data was not
fulfilled !

If the <u>number of patterns</u> n is less, equal, or only slightly greater than the <u>number of dimensions</u> d, then a linear separability may be found even for a random class assignement to the pattern points. Figure 10 shows this problem for a 2-dimensional pattern space.

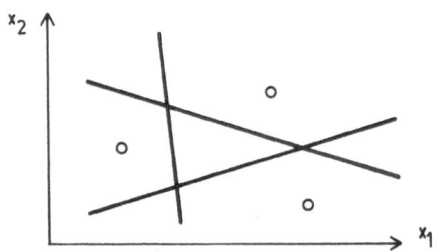

FIGURE 10. The problem of the n/d-ratio for a trivial case. Number of patterns n = 3, number of dimensions d = 2. All 3 pattern points may be arbitrarily attached to class 1 or class 2; in any case a straight line ("decision plane") is obtained that separates the classes correctly.

The following rule for the ratio n/d is now widely accepted and should be satisfied in all applications of pattern recognition :

$$n/d \geq 3 \qquad \text{is a minimum requirement}$$
$$n/d > 10 \qquad \text{is desirable} \tag{3}$$

n : number of patterns
d : number of independent features (dimensions)

Multicategory classifications require even greater values for n/d. If n/d is less than 3 for a binary classification the statistical significance of a decision plane is doubtful. A more detailed discussion of this problem is given in Chapter 10.4.

Statistical significance of a decision plane and an objective mathematical evaluation of the classifier is essential but not necessarily sufficient for successful applications of a classifer. If the relationship between features and class membership is interpreted in terms of chemical parameters one has to be very cautious and always to remember the fundamental rule "<u>correlation does not necessarily imply causation</u>".

1.7. Applications of Pattern Recognition

The general concept of pattern recognition is applicable to various classification problems in science and technology. Table 1 gives a short list of current pattern recognition applications.

TABLE 1. Some applications of pattern recognition methods [232, 370, 385]

- Recognition of printed or handwritten characters
- Analysis of photographs
- Sonar detection and classification
- Interpretation of seismic signals
- Speech recognition
- Fingerprint identification
- Interpretation of clinical data
- Medical diagnosis
- Classification in biology
- Interpretation of chemical data
- Drug design
- Quality control
- Materials analysis
- Archaeological siteing

Only a few further details can be given here. Classification of printed and written characters is probably the most frequent application of pattern recognition. Density measurements at certain screen-points are used to derive features which are suitable to recognize characters. Several methods for the recognition of shapes, figures, faces, and gestalts of objects in pictures have been developed. Aerial photographs are interpreted by using spectral densities at selected wavelengths to discriminate between different agricultural regions. Classification of chromosomes in micro-photographs has been automated by pattern recognition methods.

Analytical results of clinical chemistry or neurobiological signals are used as features to characterize diseases. Pattern recognition classifiers can be helpful in medical diagnosis.

Classification of organisms according to their relationship or

resemblance on the basis of numerical values of their characteristics
is used in <u>biology</u> (numerical taxonomy, chemical taxonomy) [272].

Applications of pattern recognition methods in <u>chemistry</u> are described in Chapters 12 to 20. Main topics in this field are: spectral
data analysis, prediction of properties from molecular structures, and
classification of materials.

Each of these disciplines has adapted the basic methods of pattern
recognition to its own specific requirements. Some authors doubt, whether
it is possible and wise to speak about pattern recognition as a single
discipline [404]. Description of pattern recognition methods in Chapters
2 to 11 will be done from the view of chemical applications.

1.8. Literature

1.8.1. General Pattern Recognition

An immense number of books and papers dealing with pattern recognition has been published by mathematicians, statisticians and other
"professional pattern recognizers". Nevertheless, it is hard to find
introductory text-books which make pattern recognition clear to a chemist who has no experience in this field. A very subjective selection
of references is given below.

<u>Relatively easy to read:</u>

- Arkadjew, Brawermann: 1966, in German [369]
- Batchelor: 1974 [370]
- Duda, Hart: 1973 [376]
- Meisel: 1972 [389]
- Meyer-Brötz, Schürmann: 1970, in German [391]
- Minsky: 1961, paper [392]
- Nilsson: 1965 [396]
- Steinhagen, Fuchs: 1976, in German [402]
- Tou, Gonzales: 1974 [403]

Other books and papers:

- Andrews: 1972 [368]
- Batchelor: 1978 [371]
- Feigenbaum, Feldman: 1963 [378]
- Fukunaga: 1972 [380]
- Mendel, Fu: 1970 [390]
- Minsky, Papert: 1969 [393]
- Nagy: 1968, paper [394]
- Niemann: 1974, in German [395]
- Patrick: 1972 [397]
- Rosen: 1967, paper [400]
- Young, Calvert: 1974 [407]

1.8.2. Pattern Recognition from the Chemist's Point of View

Numerous reviews have been written about chemical applications of pattern recognition. Many of them include rather easy to read introductions to general pattern recognition methods. At least some of these papers and books should be read by beginners of pattern recognition in chemistry.

Highly recommendable reviews (alphabetically ordered):

- Harper, Duewer, Kowalski, Fasching: 1977 [101]
- Isenhour, Jurs: 1971 [115]
- Isenhour, Jurs: 1973 [117]
- Isenhour, Kowalski, Jurs: 1974 [118]
- Jurs: 1974 [125]
- Jurs, Isenhour: Chemical Applications of Pattern Recognition, 1975,
                 book [128]
- Kowalski: 1974 [148]
- Kowalski: 1975 [149]
- Mc Gill, Kowalski: 1977 [199]
- Stuper, Brugger, Jurs: Computer Assisted Studies of Chemical Structure
                         and Biological Function, 1979, book [288]
- Wilkins, Jurs: 1978 [335]

Several excellent reviews of pattern recognition methods have been written for **mass spectroscopists**. These papers also include general overviews of pattern recognition:

- Chapman: Computers in Mass Spectrometry, 1978, book [39]
- Ward: 1971 [323]
- Ward: 1973 [324]
- Mellon: 1975 [206]
- Mellon: 1977 [207]
- Mellon: 1979 [208]

Further reviews about pattern recognition in chemistry (alphabetically ordered):

- Bender: 1973 [15]
- Carlson, Bender, Pritchard: 1975 [37]
- Clerc: 1973 [46]
- Clerc: 1977 [48]
- Coomans, Massart, Kaufman: 1979 [56]
- Currie, Filliben, De Voe: 1972 [61]
- Isenhour, Jurs: 1972 [116]
- Jurs: 1975 [126]
- Kowalski: 1977 [151]
- Kowalski, Bender: 1975 [157]
- Massart, Dijkstra, Kaufman: 1978 [189]
- Pesyna: 1975 [224]
- Pratt, Moore, Parsons, Anderson: 1978 [232]
- Shoenfeld, De Voe: 1976 [264]
- Sjöström, Kowalski: 1979 [267]
- Varmuza: 1974 [306]
- Varmuza: 1979 [309]
- Wilkins: 1977 [332]
- Ziegler: 1973 [361]

Due to the widespread applications of pattern recognition in chemistry the papers are scattered over many journals. The **journals** most frequently used in the past are listed below:

- Anal. Chem.
- Anal. Chim. Acta
- J. Am. Chem. Soc.

- J. Chem. Inf. Comput. Sci.
- J. Med. Chem.
- Appl. Spectrosc.
- J. Chromatogr. Sci.
- Technometrics
- Fresenius Z. Anal. Chem.

2. Computation of Binary Classifiers

2.1. Classification by Distance Measurements to Centres of Gravity

2.1.1. Principle

If all pattern points of a certain class form a compact cluster in the pattern space, then this class can often be well represented by the centre of gravity (centroid) of the cluster. The centre of gravity is used as a prototype (template) of that class. An unknown pattern is classified into that class which is associated with the nearest centre of gravity. In other words: the symmetry plane between the two centres of gravity is used as a decision plane (Figure 11).

This classification method is simple even for many dimensions and large training sets. Unfortunately, in many practical problems, the

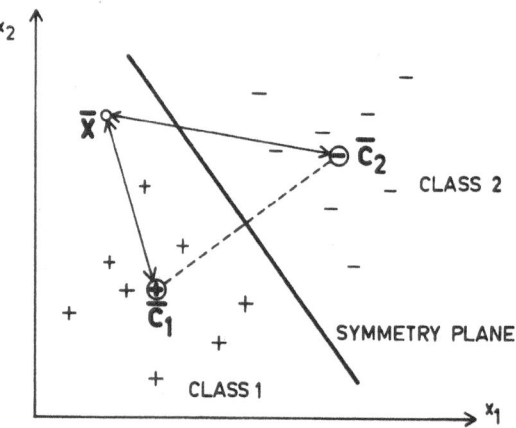

FIGURE 11. Both classes form compact clusters and can be represented by the centres of gravity $\bar{c}_1$ and $\bar{c}_2$. The unknown $\bar{x}$ is classified to belong to class 1 because the distance to $\bar{c}_1$ is shorter than that to $\bar{c}_2$. The symmetry plane can be used as a decision plane.

clusters are not tight enough to be used for this method. Classification
by distance measurements to prototypes may be used as a fist economical
approach to the application of pattern recognition methods.

2.1.2. Centres of Gravity in a d-Dimensional Space

The coordinates c_1, c_2, ... c_d of the centre of gravity $\bar{c}$ in a
d-dimensional hyperspace are calculated in the same way as for 2 dimensions. Each coordinate c_i is the arithmetic average of the components x_j
summed over all patterns j (j = 1...n) of a distinct class. Thus the
centre of gravity is the mean of all patterns belonging to the same class.

$$c_i = \frac{1}{n} \sum_{j=1}^{n} x_{ij} \qquad \text{for all dimensions i = 1...d} \qquad (4)$$

c_i : component (coordinate) i of the centre of gravity
n   : number of patterns in the class under consideration
x_{ij} : component i of pattern with number j

2.1.3. Classification by Distance Measurements

For distance measurements between two points in the d-dimensional hyperspace usually the Euclidean distance D is used. D is calculated
in the same way as for 2 or 3 dimensions by a simple extension of the
formula to d dimensions.

$$D = \sqrt{(x_1-c_1)^2 + (x_2-c_2)^2 + ... + (x_d-c_d)^2} \qquad (5)$$

$$= \sqrt{\sum_{i=1}^{d} (x_i - c_i)^2}$$

D : Euclidean distance between centre of gravity $\bar{c}(c_1, c_2, ... c_d)$
and pattern point $\bar{x}(x_1, x_2, ... x_d)$

Other distance measurements are described in Chapter 2.1.8. For simplicity, D^2 is often used instead of D; each feature i of the pattern
contributes to the squared distance by the term $(x_i - c_i)^2$.

Instead of the squared Euclidean distance D^2 an easier to calculate
distance measure $(D*)^2$ can be used which is proportional to D^2 [325].
Equation (5) can be written as

$$D^2 = \sum_{i=1}^{d} x_i^2 - 2 \sum_{i=1}^{d} x_i c_i + \sum_{i=1}^{d} c_i^2 \tag{6}$$

The first term is independent of the location of the centre of gravity
$\bar{c}$; the last term is independent of the pattern vector $\bar{x}$:

$$(D*)^2 = - \bar{x}.\bar{c} + a \qquad \text{with} \quad a = 1/2\ (\bar{c}.\bar{c}) \tag{7}$$

For the classification of unknowns a decision criterion y is defined

$$y = \Delta D = D_2 - D_1 \qquad \text{with} \quad y > 0 \ldots \text{class 1}$$
$$y < 0 \ldots \text{class 2} \tag{8}$$

D_1 : distance to centre of gravity $\bar{c}_1$ (class 1)
D_2 : distance to centre of gravity $\bar{c}_2$ (class 2)

Other formulas for the decision criterion like

$$y_2 = D_2^2 - D_1^2 \tag{9}$$

or relative differences

$$y_3 = \frac{D_2^2 - D_1^2}{D_2^2 + D_1^2} \qquad \text{or} \qquad y_4 = \frac{D_2 - D_1}{D_2 + D_1} \tag{10}$$

are possible but have no advantages [248]. Notice that the classification
by equations (8) - (10) can be easily extended to multicategory problems.

2.1.4. Classification by the Symmetry Plane

Instead of the calculation of two distances (to both centres of gravity, Chapter 2.1.3.) an equivalent but more economical procedure may be used. Two centres of gravity $\bar{c}_1$ and $\bar{c}_2$ define the symmetry plane and a corresponding weight vector $\bar{w}$ which can be used as a linear classifier. Classification with $\bar{w}$ only requires the calculation of one scalar product. Only about half of the parameters must be stored for the classifier in comparison with the use of two centres of gravity [391].

The squared Euclidean distances D_1^2 and D_2^2 between a pattern vector $\bar{x}(x_1, x_2, \ldots x_d)$ and the centres of gravity $\bar{c}_1(c_{11}, c_{12}, \ldots c_{1d})$ and $\bar{c}_2(c_{21}, c_{22}, \ldots c_{2d})$ are given by equation (5) :

$$D_1^2 = \sum_{i=1}^{d} (x_i - c_{1i})^2 \tag{11}$$

$$D_2^2 = \sum_{i=1}^{d} (x_i - c_{2i})^2 \tag{12}$$

The decision criterion $y_2 = D_2^2 - D_1^2$ (equation (9)) gives with equations (11) and (12)

$$y_2 = 2 \sum x_i(c_{1i} - c_{2i}) + \sum (c_{2i})^2 - \sum (c_{1i})^2 \tag{13}$$

Division by 2 simplifies the equation but does not affect the sign of the decision criterion.

$$y_2 = \sum_{i=1}^{d} x_i(c_{1i} - c_{2i}) + \frac{1}{2}\left[\sum_{i=1}^{d} (c_{2i})^2 - \sum_{i=1}^{d} (c_{1i})^2 \right] \tag{14}$$

This equation is compared with equation (15) which is used for classification by a weight vector $\bar{w}$.

$$s = \bar{x}.\bar{w} = \sum_{i=1}^{d} x_i w_i \tag{15}$$

A simple scalar product calculation can be used for classification if all patterns are augmented by an additional component x_{d+1} of constant value and the weight vector is also augmented by a component w_{d+1}. Comparison of equations (14) and (15) shows: If all pattern vectors $\bar{x}$ are augmented by additional components $x_{d+1} = 1$, the components w_1, w_2, ... w_d of the weight vector are given by

$$w_i = c_{1i} - c_{2i} \qquad \text{for} \quad i = 1, 2, \ldots d$$

$$\tag{16}$$

$$w_{d+1} = \frac{1}{2} \left[\sum_{i=1}^{d} (c_{2i})^2 - \sum_{i=1}^{d} (c_{1i})^2 \right]$$

The sign of the scalar product $\bar{x}.\bar{w}$ indicates the class to which a pattern $\bar{x}$ belongs (positive for class 1, negative for class 2, zero for pattern points that lie in the symmetry plane).

2.1.5. Classification by Mean Vectors

The mean vector of a class is defined by the centre of gravity. For a binary classification of an unknown pattern vector $\bar{x}$ the scalar products with both mean vectors are calculated :

$$s_1 = \bar{c}_1.\bar{x} = \sum_{i=1}^{d} c_{1i} x_i$$

$$\tag{17}$$

$$s_2 = \bar{c}_2.\bar{x} = \sum_{i=1}^{d} c_{2i} x_i$$

$\bar{x}$ is assigned to that class which gave the larger scalar product. Proper use of this classification method requires a normalization of all vectors to a constant length as shown in Figure 12 [59].

This method can be used for multicategory classifications; application to binary encoded infrared spectra gave however poorer results than distance measurements to centres of gravity [356].

A modification of this classification methods considers the a priori probabilities of the classes [133, 135].

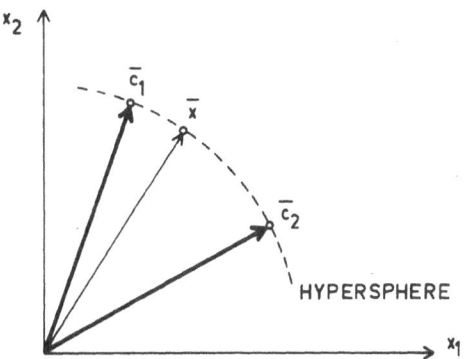

FIGURE 12. The unknown $\bar{x}$ is classified by computing scalar products with both mean vectors $\bar{c}_1$ and $\bar{c}_2$. All vectors are normalized to a constant length (all pattern vertices lie on a hypersphere). Because $\bar{c}_1 \cdot \bar{x}$ is larger than $\bar{c}_2 \cdot \bar{x}$, $\bar{x}$ is classified as member of class 1.

2.1.6. Evaluation

It is reasonable to use the same patterns for classifier development and for testing because only mean values of the patterns are important for distance measurements to centres of gravity. If the number of patterns is large enough the position of the centre of gravity does not change significantly if only a part of the training set is considered [250].

2.1.7. Projection of Pattern Points on a Hypersphere

If the pattern vectors are normalized to a fixed length r all pattern points lie on a d-dimensional sphere with radius r (Figure 13).

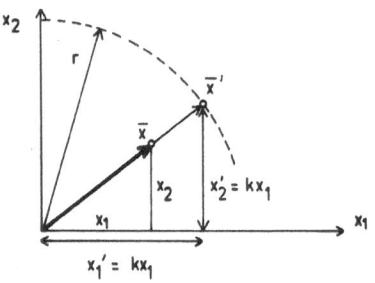

FIGURE 13. Projection of a pattern point $\bar{x}$ on a circle (d-dimensional sphere).

This normalization requires the multiplication of all vector components x_i by a factor k.

$$x_i' = k\, x_i \qquad \text{for all dimensions } i$$

(18)

$$k = r \Big/ \sqrt{\sum_{i=1}^{d} x_i^2}$$

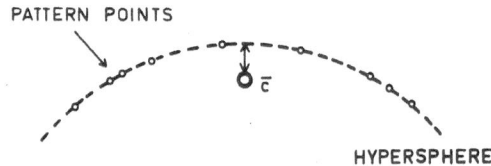

FIGURE 14. The distance between the centre of gravity $\bar{c}$ and the surface of the sphere is a criterion for the tightness of the cluster. All pattern points have been normalized to constant length.

If all pattern points lie on the surface of the sphere, then the centre of gravity of a cluster lies inside the sphere (Figure 14). The distance between the centre of gravity and the surface is a measure of the compactness of the cluster [54].

2.1.8. Distance Measurements in Pattern Space (Overview)

In this Chapter some often used distance measurements in a d-dimensional space are summarized.

Let $\bar{x}(x_1, x_2, \ldots x_d)$, $\bar{z}(z_1, z_2, \ldots z_d)$, and $\bar{u}(u_1, u_2, \ldots u_d)$ be three points in a d-dimensional space. Any function $D(\bar{x}, \bar{z})$ satisfying

$$D(\bar{x}, \bar{x}) = 0 \tag{19}$$
$$D(\bar{x}, \bar{z}) > 0 \quad \text{for} \quad \bar{x} \neq \bar{z} \tag{20}$$
$$D(\bar{x}, \bar{z}) = D(\bar{z}, \bar{x}) \tag{21}$$
$$D(\bar{x}, \bar{u}) + D(\bar{u}, \bar{z}) \geq D(\bar{x}, \bar{z}) \tag{22}$$

has the quality of a distance. Definition of a distance is therefore arbitrary within these conditions [370, 389].

The most frequently used distance measurements are derived from the general <u>Minkowski distance</u>

$$D_{Minkowski} = \left[\sum_{i=1}^{d} (x_i - z_i)^k \right]^{1/k} \tag{23}$$

k = 2 gives the <u>Euclidean distance</u> which is most often used

$$D_{Euclidean} = \left[\sum_{i=1}^{d} (x_i - z_i)^2 \right]^{1/2} \tag{24}$$

k = 1 gives the "<u>city block distance</u>" (<u>Manhatten distance</u>, in analogy to the shortest way between two points in a city with rectangular streets, Figure 15).

$$D_{city\ block} = \sum_{i=1}^{d} |x_i - z_i| \tag{25}$$

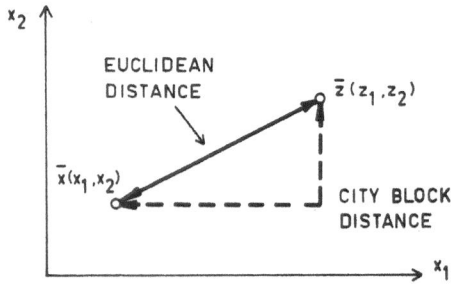

FIGURE 15. Euclidean distance and city block distance.

If all features x_i and z_i are <u>binary encoded</u> (only the values '0' or '1' are allowed) the city block distance is often called <u>Hamming distance</u>. The Hamming distance is equivalent to the number of features which are different in $\bar{x}$ and $\bar{z}$. Application of logical exclusive-or (XOR) commands (Table 2) in computer programs permits rapid calculation of the Hamming distance.

$$D_{Hamming} = \sum_{i=1}^{d} XOR(x_i, z_i) \qquad (26)$$

The Hamming distance can also be written in another form [353]:

$$D_{Hamming} = b_x + b_z - 2b_{xz} \qquad (27)$$

b_x : number of binary '1's in pattern $\bar{x}$
b_z : number of binary '1's in pattern $\bar{z}$
b_{xz} : number of features with a '1' in both patterns $\bar{x}$ and $\bar{z}$

The Hamming distance has some disadvantages for vectors with only a few components containing '1's. An alternative is the <u>Tanimoto distance</u> which is a normalized Hamming distance [398].

$$D_{Tanimoto} = \frac{D_{Hamming}}{b_x + b_z - b_{xz}} = \frac{AND(x_i, z_i)}{OR(x_i, z_i)} \qquad (28)$$

The denominator in equation (28) is the number of vector components which contain a '1' at least in one of the patterns (logical or); The numerator is the number of vector components with a '1' in both vectors (logical and, Table 2). Application of the Tanimoto distance to the classification of infrared spectra yielded better results than the Hamming distance [353].

TABLE 2. Logical functions AND, OR, XOR (and, or, exclusive or).

x_i	z_i	$AND(x_i, z_i)$	$OR(x_i, z_i)$	$XOR(x_i, z_i)$
0	0	0	0	0
0	1	0	1	1
1	0	0	1	1
1	1	1	1	0

If the number d of vector components is not the same in all distance calculations a normalized distance $D/\sqrt{d}$ is useful [175, 187].

2.1.9. Distance Measurements with Weighted Features (Generalized Distances)

The Euclidean distance (equation (5)) can be generalized if each part of the sum (each feature i) is given a weight g_{im}. The weights depend on the feature number i and on the class m.

$$D_{gm}^2 = \sum_{i=1}^{d} g_{im} (x_i - c_{im})^2 \tag{29}$$

D_{gm}^2 : generalized squared distance between pattern point $\bar{x}(x_1, x_2, \ldots x_d)$ and centre of gravity $\bar{c}_m(c_{1m}, c_{2m}, \ldots c_{dm})$ that belongs to class m

The generalized distance can be interpreted as a linear transformation of the coordinates; new coordinates x_i' of a pattern point are given by equation (30).

$$x_i' = x_i \sqrt{g_{im}} \tag{30}$$

In the new pattern space the Euclidean distance is used again.

A first approach to the calculation of the weights g_{im} is based on the minimization of the averaged squared distance within each class [391: p.42].

$$g_{im} = \frac{1}{v_{im}} \prod_{j=1}^{d} v_{jm}^{1/d} \tag{31}$$

v_{im} : variance of feature i over all n_m patterns of class m

$$v_{im} = \frac{1}{n_m - 1} \sum_{j=1}^{n_m} (x_{ij} - c_{im})^2 \tag{32}$$

The weight is primarily proportional to the inverted variance.

For a binary decision a criterion y (equation (9)) can be used.

$$y = D_1^2 - D_2^2 \qquad \begin{array}{l} y < 0 \ \dots \ \text{class 1} \\ y > 0 \ \dots \ \text{class 2} \end{array} \tag{33}$$

Insertion of equation (29) gives

$$y = \sum_{i=1}^{d} \left[g_{i1}(x_i - c_{i1})^2 - g_{i2}(x_i - c_{i2})^2 \right] \tag{34}$$

$$= \sum_{i=1}^{d} (w_i x_i^2 + z_i x_i + a_i) \tag{35}$$

Equations (34) and (35) define a classifier which is summarized in
equation (36).

$$y = \bar{w}.(\bar{x})^2 + \bar{z}.\bar{x} + a \qquad (36)$$

with : $w_i = g_{i1} - g_{i2}$

$z_i = 2 (g_{i2}c_{i2} - g_{i1}c_{i1})$

$a = \sum_{i=1}^{d} a_i = \sum_{i=1}^{d} (g_{i1}c_{i1}^2 - g_{i2}c_{i2}^2)$

$y < 0$ ... class 1
$y > 0$ ... class 2

Equation (36) represents a quadratic classifier with 2d+1 parameters
($w_1, w_2, \ldots w_d, z_1, z_2, \ldots z_d, a$). In the case of binary encoded
features this method becomes linear.

In practical applications of this method difficulties arise if a
certain feature is zero for all members of a class. Then, the corresponding variance is zero and the weight for this feature becomes infini-
tive. The weights for all other features (with non-zero variance)
become zero. A properly chosen minimum value for the variances may overcome this problem.

Another approach to the calculation of the weights considers all
classes simultaneously. A measure of importance has to be defined which
reflects minimum intraclass distances (within a class) and maximum inter-
class distances (between classes) [2, 391: p.48].

2.1.10. Chemical Applications

Classification by distance measurements to centres of gravity
requires compact clusters which are often not present in chemical applications. Usually, other classification methods give better results.
Nevertheless, this simple and evident method serves as a standard for
comparisons with more sophisticated pattern recognition methods.

An extensive application was reported for the recognition of
the structures of steroid molecules from low resolution mass
spectra [248, 250, 315]. Classification by generalized distances (quadratic classifier, equation (36)) gave significantly better results than
classification by Euclidean distances [248].

Other applications concern infrared spectra [133, 135, 353, 356]
and mass spectra of various chemical classes [59].

2.2. Learning Machine

2.2.1. Principle

A learning machine can be generally defined as "any device whose
actions are influenced by past experience" [396]. Learning machines
which are used in chemical pattern recognition problems are rather simple
devices within this definition and have almost no similarity to human
learning. The learning machine approach was one of the first [129] and
probably the most frequently used pattern recognition method in chemistry. It seems that the very promising – but also misleading – name of the
method has initiated in the past much of the enthusiasm and opposition
which are often connected with applications of pattern recognition in
chemistry. The learning machine – as it was used in chemical applications
– is in fact not much more than a simple iterative procedure to find a
decision plane situated between two clusters in a d-dimensional space.
A less pretentious synonym for this method is "classification by
adaptive (linear, binary) classifiers".

The training ("learning") starts with an arbitrary selected plane
passing through the origin. The plane is represented by the decision
vector $\bar{w}$ (Chapter 1.3) which starts at the origin and is orthogonal
to the plane. A representative set of patterns is used as a training set
to adjust the decision vector until it classifies all members of the
training set correctly. The patterns $\bar{x}_j$ of the training set are presented to the classifier one after another. When a correct classification
is made (that means the scalar product $\bar{x}_j.\bar{w}$ has the correct sign)
no action is taken. Whenever the classification is incorrect the decision

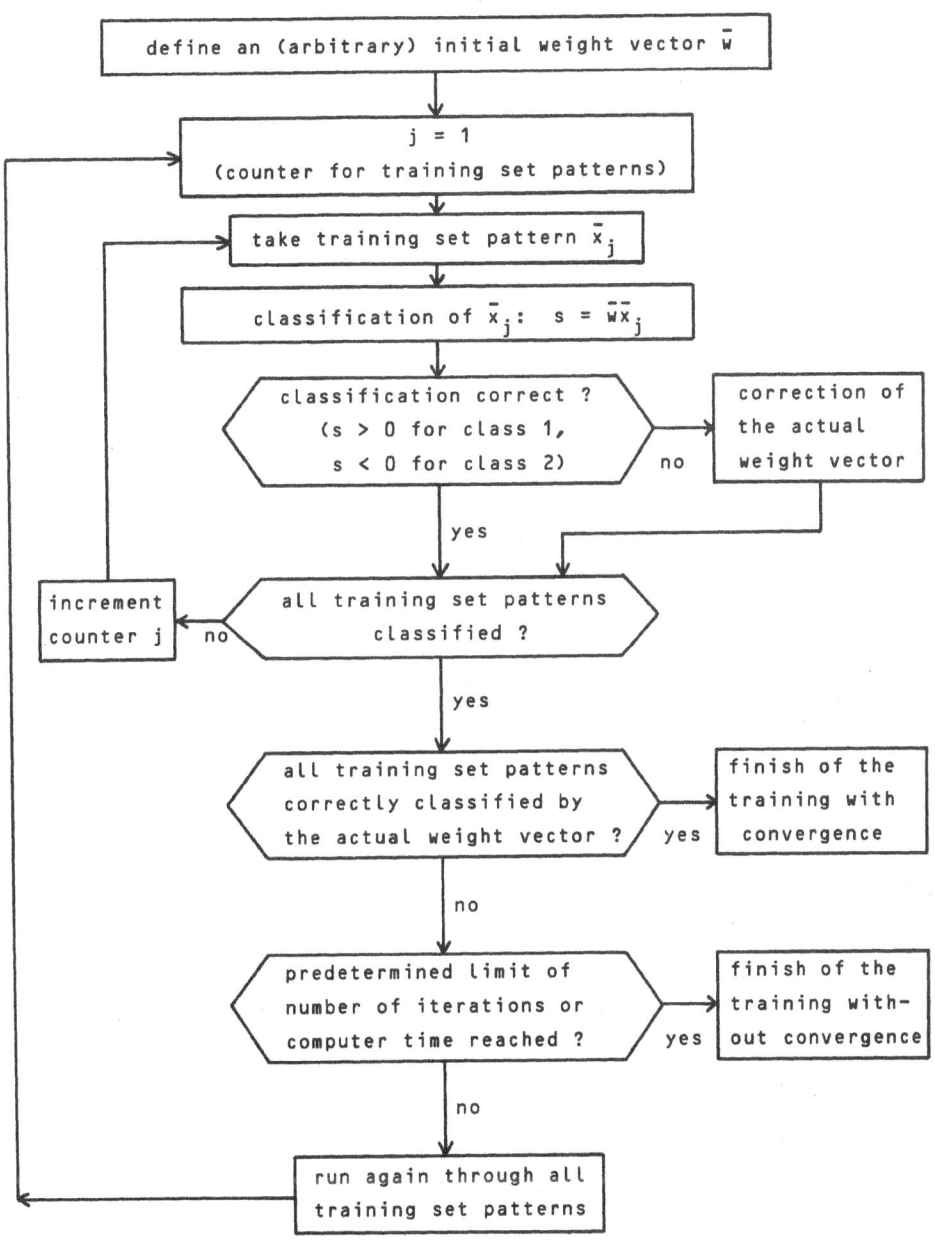

FIGURE 16. Scheme of the learning machine: Training of an adaptive, binary, linear classifier.

vector is adjusted so that the new decision vector will correctly classify the previously misclassified pattern. The classifier "learns" by a negative feedback. This process continues until all patterns of the training set are correctly classified (Figure 16). It was shown by mathematicians that this training procedure will converge and find a separating plane if one exists at all [391, 396]. Unfortunately, the theory does not say how many iterations (corrections of the decision vector) are necessary to find a decision plane that will correctly classify all members of the training set. In practice, the training is stopped at a predetermined number of iterations or at a certain computer time limit.

Characteristic values of the training process are the <u>convergence rate</u> (number of iterations necessary) and the <u>recognition rate</u> (percent correctly classified patterns of the training set by the final decision vector). The evaluation of the decision vector must be performed with a second set of patterns (prediction set) which were not used for the training. Although the learning machine approach is fascinating in its idea one must not overlook some constraints and man traps of this method (Chapter 2.2.5).

Various types of learning machines are described in the literature. The book "Learning Machines" written by Nilsson [396] was probably the stimulant for extensive applications of this method in chemistry.

The learning machine in connection with chemical applications was described in detail by several authors [alphabetical: 37, 39, 115, 117, 118, 128, 129, 131, 163].

2.2.2. Initial Weight Vector

In principle, an arbitrary selected decision vector - often called "weight vector" - may be used to start the training. But it was shown that the initial weight vector influences both the convergence rate of training and the final position [131]. In many works all components w_i of the weight vector $\bar{w}(w_1, w_2, \dots w_d)$ are initialized to the value +1 or -1. Wilkins et.al. [336] found for a training set in which class 1 was more populous than class 2 a greater convergence rate if w_i is initiated with +1 (instead of -1) and vice versa.

An initial weight vector with unit length and components

$$w_i = 1 / \sqrt{d} \qquad (i = 1, 2, \dots d) \qquad (37)$$

was suggested [44, 94]. An initial weight vector with all components
set to zero - except the last, additional one (Chapter 1.3) - was also
proposed [44, 168].

The convergence rate of the training is significantly increased if
the initial weight vector already contains information about the clusters
to be separated. A suitable way is to use the symmetry plane between the
two centres of gravity for the beginning of the training (see Chapter
2.1.4, equation (16)). Additional use of the a priori probabilities of
class 1 and 2 still improves the initial position of the decision plane.
A significant increase of the convergence rate by using this method was
reported [94, 128]. For a training set of 80 patterns and 5 dimensions
only 3 feedbacks instead of 254 were necessary. A simpler approach to
increase the convergence rate is to start the training with the difference of the class mean vectors (Chapter 2.1.5), [53, 184].

Lidell and Jurs [167] proposed an initial weight vector with components

$$w_i = \frac{n_{i1}}{n_1} - \frac{n_{i2}}{n_2} \tag{38}$$

n_1, n_2 : number of patterns in class 1 (resp. class 2)
n_{i1}, n_{i2} : number of patterns in class 1 (resp. class 2) with only
          positiv values of feature i

Whenever possible a pre-trained initial weight vector should be used
in order to increase the convergence rate [311].

As described in Chapter 1.3 all original pattern vectors must be
augmented by a constant additional component to obtain a decision plane
which passes through the origin. If this additional component is much
larger than an average pattern component the convergence rate may decrease extremely [322].

2.2.3. Correction of the Weight Vector

If a pattern is misclassified during the training phase the weight
vector has to be corrected (adapted). An evident method is to change the
weight vector in a way that it classifies the previous misclassified
pattern correctly.

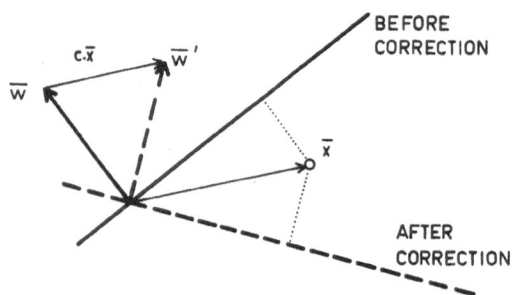

FIGURE 17. Correction of the weight vector $\bar{w}$ which gave a wrong classification for pattern $\bar{x}$ ($\bar{x}$ is considered as a member of class 1 and
should give a positive scalar product). The new weight vector $\bar{w}'$ classifies $\bar{x}$ correctly.

Figure 17 shows that a new, corrected weight vector $\bar{w}'$ can be obtained by adding a fraction of the misclassified pattern vector $\bar{x}$ to the
actual weight vector $\bar{w}$

$$\bar{w}' = \bar{w} + c.\bar{x} \tag{39}$$

Widely used is a <u>reflexion</u> of the old decision plane on the misclassified
pattern. After this correction the distance between $\bar{x}$ and the decision
plane is the same as before but $\bar{x}$ lies at the correct side of the plane.
The same consideration is valid for the scalar product s before correction and s' after correction.

$$s' = -s \tag{40}$$

$$\bar{w}'.\bar{x} = -\bar{w}.\bar{x} \tag{41}$$

Insertion of equation (39) gives an equation for the correction factor c

$$(\bar{w} + c.\bar{x}).\bar{x} = -\bar{w}.\bar{x} \tag{42}$$

Insertion of $s' = \bar{w}'.\bar{x}$ and equation (39) gives

$$c = -\frac{2s}{\bar{x}.\bar{x}} \tag{43}$$

If a training set pattern $\bar{x}$ gives a scalar product with a wrong sign then c is calculated by equation (43) and the weight vector $\bar{w}$ is corrected by equation (39). $\bar{x}.\bar{x}$ corresponds to the sum of the squared components of pattern $\bar{x}$.

Extensive studies by Jurs et. al. [131] with mass spectra showed that a reflexion of the decision plane yields the highest convergence rate. Other methods of weight vector correction and a method with a small positive feedback upon correct classification were investigated by the same authors but did not give better results.

A "fractional correction method" was described by Lidell and Jurs [168]. In this method the weight vector is split during the training into two components, one for class 1 and one for class 2. This method is less sensitive to the ordering of patterns in the training set than the usual procedure. Another sophisticated method called "MAX" was described by the same authors [167] and preferably uses those patterns for the training which lie close to the decision plane. Mahle and Ashley [184] reported a training method based on correlation coefficients.

2.2.4. Methods of Training

The general scheme of the learning process is shown in Figure 16. A faster procedure was described by Isenhour et. al. [89, 120, 319, 322] and is called "collapsing subset procedure" (Figure 18). Patterns which are misclassified are placed in a new subset and this subset is used in the next iteration run (subset-run). If no errors occur in a subset-run the program proceeds through previous subsets [89, 322] or, in another version, through the entire training set [120, 310]. This process continues until all patterns of the training set have been correctly classified by the same weight vector or a predetermined number of iterations or computer time is reached.

Some characteristics of the convergence rate have been examined with applications in mass spectrometry [131]: Large training sets require many iterations or do not reach convergence within reasonable computer time. Preprocessing (Chapter 9) of mass spectra with a decrease of the dynamic range of the features (logarithm or square root or autoscaling of the peak heights) increased the convergence rate but did not influence the predictive abilities significantly [44, 123]. If new patterns are to be

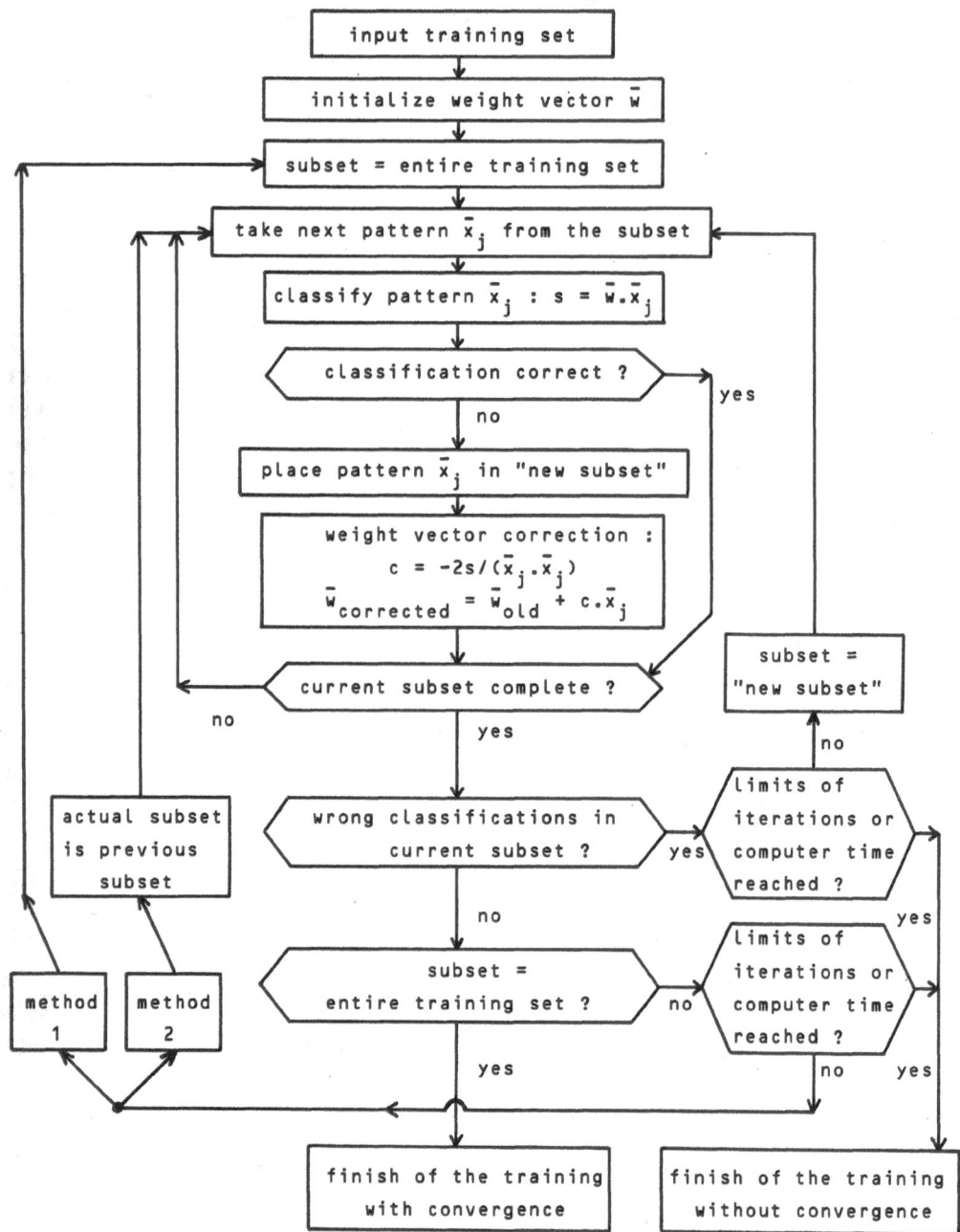

FIGURE 18. Subset training of the Learning machine. The weight vector is corrected by reflexion of the decision plane at the misclassified pattern.

included in the training set then it is necessary to start a new training
with all patterns. Jurs [123] made some successful considerations on
mass spectra whether the training set could be represented by a subset of
its members.

2.2.5. Restrictions

A simple learning machine as described in Figures 16 or 18 stops
the training at the moment where all members of the training set are
correctly classified. The final position of the decision plane depends
on the initial position and on the sequence of the training set patterns.
An accidental bad classifier may result as shown in Figure 19a. Especially the first large corrections during the training influence the final
position significantly [217]. The final position may also be unpleasantly
determined by atypical outliers [46] as shown in Figure 19b. Most of
these problems are avoided if a "dead zone training" is used in order to
place the decision plane in the middle between two clusters (Chapter
2.2.7).

A fundamental restriction of all these learning machines is the
aim of the learning process, namely a plane which classifies correctly
the whole training set. Other optimization criteria may be more useful in
certain cases [318]; see Chapters 2.3 and 2.4.

If the training set is not linearly separable (even a single member of the training set may cause it) the decision plane oscillates and
the training does not converge. If the training is stopped after a predetermined number of iterations a randomly bad classifier may result.
Ritter, Lowry, Isenhour, and Wilkins [242, 243] proposed an "average
subset learning machine" to overcome this difficulty: training is done
with several small linearly separable subsets of the training set and all
resulting weight vectors are summed to an averaged final weight vector.
Mac Donald [180] tested and stored the weight vectors after each iteration during the training; if the training is stopped without convergence, the best of all calculated weight vectors can be selected.

It was shown that the predictive ability may change considerably
during the training [131].

Another training procedure was proposed by Mathews [192] to overcome linear inseparability. If the training does not converge within 20
iterations, then atypical patterns of class 2 are transferred from the

training set to the prediction set. If the test of the weight vector with
the prediction set gives an unsatisfactory performance of the classifier,
then misclassified patterns of class 1 are transferred from the prediction set to the training set and the training is repeated.

Addition of class-independent "noise-features" to all patterns had
the surprising effect of an increase of the convergence rate; however,
the predictive ability decreased significantly [327].

Training sets with a small n/d ratio (number of patterns / number of
dimensions) may cause considerable confusions. These problems are discussed in Chapters 2.2.6, 10.3.2 and 10.4.

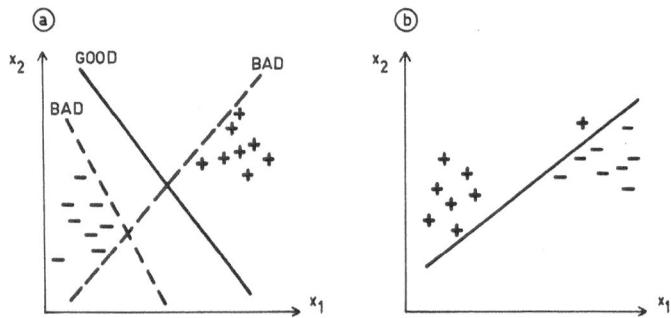

FIGURE 19. Unpleasant final positions of the decision plane after
training with a learning machine.
a) Arbitrary bad position of the decision plane.
b) Atypical outliers may determine the final position of the decision
   plane. Although this plane separates the clusters completely the
   result is not satisfactory.

2.2.6. Evaluation

The randomness of the final position of a decision plane - computed
with the learning machine - requires special cautions in the evaluation.
If the number n of patterns in the training set is less than 3 times the
number d of dimensions, a senseless decision plane may have been obtained.

A ratio n/d of 5 or 10 is desirable [8, 141]; see Chapters 1.6 and 10.4.
    Even a randomly partitioning of a training set into two classes may give linear separability and convergence in the training process if the portion of one class is less than 10 % [53, 94, 251, 310].
    Evaluation of the classifier with a sufficiently large prediction set is absolutely necessary [94, 327]. For small data sets the "leave-one-out-method" was proposed for an objective evaluation [44]. The predictive ability is usually higher for that class which is more frequent in the training set [327].
    Learning machine experiments with random patterns have been reported by Anderson and Isenhour [8], and by Gray [94].
    Convergence of the training process (100 % recognition) does not prove a chemical meaningful relationship between patterns (e.g. spectra) and the classification problem (e.g. the presence of a certain molecular structure). This is valid for all pattern classification methods but must be emphasized especially in connection with learning machines.

2.2.7. Dead Zone Training

    If the decision plane between two clusters is given a finite thickness, then the final position of the plane will be in the middle between the clusters (Figure 20). An optimum position of the decision plane can be achieved if the thickness is enlarged stepwise until the training does not converge.
    Patterns $\bar{x}$ are considered to be correctly classified by the weight vector $\bar{w}$ if the scalar product s is

$$s = \bar{w}.\bar{x} \qquad \begin{array}{ll} s > t & \text{for class 1} \\ s < -t & \text{for class 2} \end{array} \qquad (44)$$

otherwise the weight vector must be corrected. Convergence of the training is therefore obtained if all patterns of the training set lie on the correct side of the decision plane and the dead zone is empty.
    The scalar product s is used to measure the distance between a pattern $\bar{x}$ and the decision plane. Because s not only depends on the position of $\bar{x}$ and $\bar{w}$ but also on the length of the vectors, the weight vector must be normalized to a fixed length $|\bar{w}|$ (e.g. to length 1); Chapter 2.1.7.

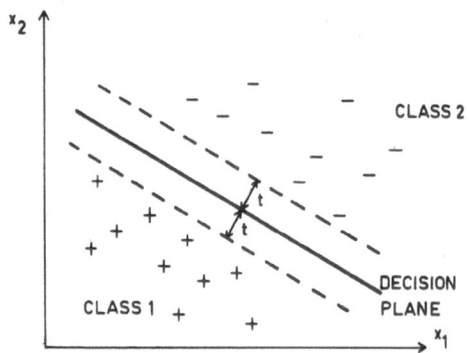

FIGURE 20. Decision plane with thickness 2t. The training process
tries to find a decision plane with no patterns within a dead zone of
±t from the plane.

$$|\bar{w}| = (\bar{w}.\bar{w})^{1/2} = \left[\sum_{i=1}^{d} w_i^2\right]^{1/2} = \left[w_1^2 + w_2^2 + \ldots + w_d^2\right]^{1/2} \qquad (45)$$

Normalization of the weight vector can be accomplished after each weight
vector correction as proposed by Jurs et. al. [123, 167, 168, 233]. Another method was proposed by Wangen, Frew, and Isenhour [320] who included the normalization into the weight vector correction procedure.

Several chemical applications of the learning machine showed that
a maximized thickness of the decision plane significantly increased the
performance of the classifier [123, 167, 168, 233, 319, 320].

If the classifier is applied to unknown patterns, the dead zone
from the training can be used as a rejection zone: Patterns which give
a scalar product between -t and +t are not classified. Chemical applications showed that the predictive ability for the remaining patterns
outside the rejection zone is increased by this method [320]. The distance to the decision plane may be used as a measure of confidence for
classifiers with a continuous response (Chapter 2.6.1.).

Another benefit of a decision plane with a definite width results
from application of a negative width to linearly inseparable classes as
shown in Figure 21. Patterns in the overlapping region (dead zone) are
excluded from classification and the learning machine gets the chance to
converge. The no-decision-region is minimized during consecutive training processes [320].

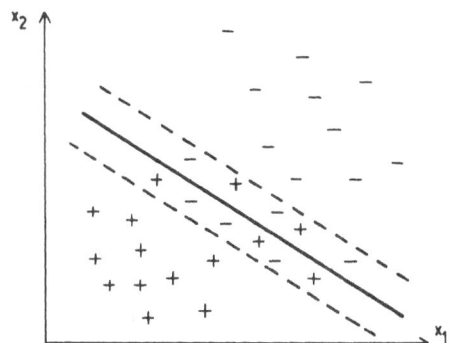

FIGURE 21. Decision plane with a "negative thickness" defines a no-
-decision region for linearly inseparable data. This trick enables the
learning machine to converge.

2.2.8. Chemical Applications

The learning machine was tested for many chemical applications of
pattern recognition, especially the interpretation of mass spectra was
investigated in detail [128]. The results are usually somewhat better
than for classification by distance measurement. Some caution is necessary with reported results in early publications because the restrictions
to this method have not always been fully considered. The learning machine is advantageously used if the clusters are linearly separable and
if the whole training set can be stored in the memory of the computer
(the patterns are used more than once and in different sequences).

In consequence of the randomness of the trained weight vector a
classification by a committee machine (Chapter 2.6.2.) or a dead zone
training is advisable [117, 123, 159, 295].

The usefulness of the learning machine method for chemical applications has been called in doubt by several authors. Typical predictive
abilities of 70 to 95 % may be insufficient for consecutive binary decisions [46]. The learning machine may probably be applicable only to
simple and very limited classification problems in chemistry [46, 119,
318]. It seems to be difficult at the moment to formulate a definite
opinion about the utility of the learning machine for chemical problems.

2.3. Linear Regression (Least-Squares Classification)

2.3.1. Principle

Classification of a d-dimensional pattern vector $\bar{x}$ by a linear classifier is performed by computing the scalar product s (Chapter 1.3, equation (2)).

$$s = \bar{w}.\bar{x} \tag{46}$$

The components w_1, w_2, ... w_d of the weight vector $\bar{w}$ are the parameters of a decision function (discrimination function) $s = s(x_1, x_2, ... x_d)$ which is used to classify $\bar{x}$ (e.g.: s is positive for class 1 and negative for class 2). The parameters of the decision function are adjusted during the training in a way that as much as possible patterns of the training set are correctly classified. This adjustment of the decision function can be carried out by application of a regression analysis [17, 118, 128, 148, 251, 318, 391].

A demand is set up about the value of the scalar product: it should be +z for all patterns belonging to class 1 and -z for all patterns belonging to class 2. A weight vector $\bar{w}$ is sought that fulfills this demand in an optimum way. A convenient approach to this optimization problem is the least-squares error method. The error e_j which is made when a pattern $\bar{x}_j$ is classified is given by the difference of the actual scalar product s_j and the desired value z_j.

$$e_j = z_j - s_j \tag{47}$$

Minimization of the sum of all squared errors in the training set

$$\sum_{j=1}^{n} (e_j)^2 \longrightarrow \text{minimum} \tag{48}$$

gives the parameters w_1, w_2, ... w_d for the decision function. A complete mathematical treatment is given in Chapter 2.3.2.

Linear regression analysis allows to find a well defined, and in some way optimum decision plane even for overlapping clusters.

Linear regression can also be used for multicategory classifications. For each class m a desired value z_m for the scalar product is demanded. Only one weight vector is necessary to discriminate several classes [17, 160].

Prediction of a continuous property (instead of discrete classes) can be done by linear regression without modifications of the method. The desired values of the scalar products are given by the numerical values of the property (or a function of them) [148].

Computation of a classifier by linear regression requires some mathematical and computational efforts because a system of d linear equations with d variables has to be solved. Application of the classifier to unknown patterns is effected by computing a scalar product in the same simple way as with adaptive classifiers (learning machine) or with distance measuring classifiers.

2.3.2. Mathematical Treatment

The aim of regression analysis is to determine the statistical relation between pattern vectors $\bar{x}_j$ and a scope value z which is used to classify a pattern. A function $s = s(\bar{x})$ that approximates z as closely as possible is therefore sought. For a binary classifier a scope value (forcing value) of $z = +1$ can be demanded for all pattern vectors of class 1 and a value of $z = -1$ for class 2.

The decision function s has to be determined so that the expected value of the squared error $E[z-s]^2$ is minimized.

$$E[z-s]^2 = \frac{1}{n} \sum_{j=1}^{n} (z_j - s_j)^2 \quad ---> \quad \text{minimum} \tag{49}$$

For this calculation a set of n pattern vectors with known class memberships is necessary. For a linear classifier the decision function s has the general form

$$
\begin{aligned}
s &= s(x_1, x_2, \ldots x_d) \\
&= w_1 x_1 + w_2 x_2 + \ldots + w_d x_d \\
&= \sum_{i=1}^{d} w_i x_i
\end{aligned}
\tag{50}
$$

The function s is determined by the parameters w_i (i = 1...d) and is a linear function of the components x_i (i = 1...d) of the pattern vectors.

For simplicity, one component with a fixed value is added to all pattern vectors in the same way as for other classification methods (Chapter 1.3.). Thus the decision plane obtained passes through the origin of the coordinate system. The total number of dimensions (including the added one) is denoted by d.

The parameters w_i of the decision function form the decision vector $\bar{w}$ which is perpendicular to the decision plane required. The scalar product of decision vector $\bar{w}$ and pattern vector $\bar{x}$ gives the classification result. By definition, positive values refer to class 1 (z = +1) and negative values to class 2 (z = -1).

In order to establish the decision vector which minimizes the sum of the squared errors as given in equation (49), $E[z-s]^2$ is differentiated with respect to w_i and the derivatives are set to zero.

$$\frac{\partial E[z-s]^2}{\partial w_i} = 0 \qquad \text{for all } w_i, \text{ with } i = 1, 2, \ldots d \qquad (51)$$

Because z is independent of w_i equation (51) gives

$$\frac{\partial E[z-s]^2}{\partial w_i} = -2 E\left[z \frac{ds}{dw_i}\right] + 2 E\left[s \frac{ds}{dw_i}\right] = 0 \qquad (52)$$

Substitution of equation (50) for s gives the following relations for a set of d partial derivatives.

$$\frac{\partial s}{\partial w_i} = \frac{d (w_1 x_1 + w_2 x_2 + \ldots w_i x_i + \ldots + w_d x_d)}{dw_i} \qquad (53)$$

$$= x_i$$

Substitution of equations (50) and (53) into equation (52) yields

$$-E[zx_i] + E\left[x_i \sum_{h=1}^{d} w_h x_h\right] = 0 \qquad \text{for } i = 1 \ldots d \qquad (54)$$

A change of summation and formation of the expected values leads to

$$\sum_{h=1}^{d} \left[w_h \ E[x_i x_h] \right] \quad = \quad E[zx_i] \qquad \text{for} \quad i = 1 \ldots d \qquad (55)$$

Relation (55) defines a system of d linear equations for d variables w_1, w_2, ... w_d. This set of equations is written in detail :

$$i=1 : \quad w_1 \ E[x_1 x_1] \ + \ w_2 \ E[x_1 x_2] \ + \ \ldots \ + \ w_d \ E[x_1 x_d] \quad = \quad E[zx_1]$$

. . .

. . .   and so on for $i = 1 \ldots d$ (56)

. . .

$$i=d : \quad w_1 \ E[x_d x_1] \ + \ w_2 \ E[x_d x_2] \ + \ \ldots \ + \ w_d \ E[x_d x_d] \quad = \quad E[zx_d]$$

Solution of this set of equations gives the desired components of the decision vector $\bar{w}$.

The expected values $E[x_i x_h]$ and $E[zx_i]$ can be estimated from a random sample (training set) of n pattern vectors $\bar{x}$ for which the class membership and thus z are known.

$$E[x_i x_h] \quad = \quad \frac{1}{n} \sum_{j=1}^{n} x_{ij} x_{hj} \qquad \text{for i and h} = 1, 2, \ldots d \qquad (57)$$

$$E[zx_i] \quad = \quad \frac{1}{n} \sum_{j=1}^{n} z_j x_{ij} \qquad \text{for} \quad i = 1, 2, \ldots d \qquad (58)$$

x_{ij} is the i-th component of pattern vector number j.

To summarize, the following steps are necessary for the calculation of a classifier by linear regression analysis :

1. Calculation of the expected values (averaged values) $E[x_i x_h]$ for all d^2 combinations of $x_i x_h$ (i = 1...d; h = 1 ... d) with equation (57).
2. Calculation of the expected values (averaged values) $E[zx_i]$ for i = 1...d; z = +1 for class 1; z = -1 for class 2 with equation (58).
3. Generation of a set with d linear equations (56).
4. Solution of the set with d linear equations and d unknowns by using standard methods of numerical mathematics.

The computational effort for steps 1 to 3 is proportional to the number of dimensions d and to the number of patterns n; step 4 is proportional to d. For modern computers of moderate size usually programs for solving a set of linear equations are available and can be executed without problems for $d \leq 100$.

2.3.3. Characteristics and Variations of the Method

Linear regression analysis is advantageous if the patterns of the training set are not linearly separable. If a training set is actually linearly separable, a classifier computed by linear regression may give poorer results than a classifier computed by the learning machine [118]. A computational advantage of linear regression is that each pattern is read (into the computer) only once and a unique decision plane is obtained. The position of the decision plane is only less sensitive to outlying pattern points. After the regression analysis, a "best line routine" may be used to find an optimum discriminating value of the scalar product which gives the best classification result for the training set [160].

If one class is more frequent in the training set than the other, the resulting classifier may recognize the more frequent class well but is unsatisfactory for the other class [157, 229]. To overcome this problem it is useful to construct a fictitious sample with equal numbers of patterns in both classes [248, 251]. For this purpose, the expected values E in equations (57) and (58) are split into two class-dependent parts E_1 (expected value for class 1) and E_2 (for class 2). If p(1) is defined as the probability of class 1 in the training set, then

$$E = p(1) E_1 + \{1 - p(1)\} E_2 \qquad (59)$$

To obtain the expected value E for a fictitious sample with equal probabilities of both classes, p(1) is given the value 0.5. Instead of equations (57) and (58) the following equations are used to determine the coefficients of the set of linear equations (56).

$$E[x_i x_h] = \frac{0.5}{n_1} \sum_{j=1}^{n_1} x_{ij} x_{hj} + \frac{0.5}{n_2} \sum_{j=1}^{n_2} x_{ij} x_{hj} \qquad (60)$$

$$\text{for i and h} = 1 \ldots d$$

$$E[zx_i] = \frac{0.5}{n_1} \sum_{j=1}^{n_1} z_j x_{ij} + \frac{0.5}{n_2} \sum_{j=1}^{n_2} z_j x_{ij} \qquad (61)$$

$$\text{for } i = 1 \ldots d$$

n_1 and n_2 are the numbers of patterns belonging to class 1 and 2.

Pietrantonio and Jurs [229] introduced the hyperbolic tangent function (tanh) into linear regression in chemical applications of pattern
recognition. The hyperbolic function is well suited for pattern dichoto-
mizers because it is positive for all positive values of the independent
variable and negative for all negative values (Figure 22).

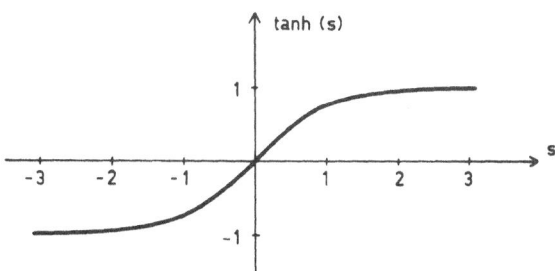

FIGURE 22. The hyperbolic tangent function tanh(s)

Instead of the scalar product s in equation (49) the tanh(s) is
used and, according to the least-squares principle, the function to be
minimized becomes

$$\frac{1}{n} \sum_{j=1}^{n} \{z_j - \tanh(s_j)\}^2 \quad \text{---}> \quad \text{minimum} \qquad (62)$$

The weight vector can be found by an iterative procedure [118, 128, 229,
260, 303].

Other methods of linear regression for chemical applications of
pattern recognition have been used by Bos and Jasink [21], Kowalski and
Reilly [161], and Volkmann [318].

2.3.4. Chemical Applications

Linear regression analysis for pattern recognition in chemistry was primarily used for automatic interpretation of mass spectra. Comparisons of linear regression and learning machine gave different results. Classification of oxygen presence and number of oxygen atoms in a molecule by linear regression yielded poorer results than a learning machine [160] but in the recognition of the molecular structures of steroids the linear regression showed significant better results [251]. Linear regression required much less features to obtain the same classification result as a learning machine [229, 260].

Although the linear regression sometimes gave a lower recognition rate and predictive ability for linearly separable data than a learning machine, linear regression seems to be a promising method. The position of the decision plane is very satisfactory, even for overlapping clusters; and its orientation is not greatly affected by a few patterns in extreme positions.

2.4. Simplex Optimization of Classifiers

2.4.1. Principle

The search for a decision plane may be considered as an optimization problem. The decision vector $\bar{w}(w_1, w_2, \ldots w_d)$ is defined by d components (Chapter 1.3.). The optimization problem is to find a combination of values for all components w_i that best separates the clusters. The "response" which has to be maximized during the optimization may be the recognition rate (percentage of correctly classified patterns from the training set). The response-surface is a (d+1)-dimensional mountain, located above a d-dimensional plane in a coordinate system with the axes $w_1, w_2, \ldots w_d$. Optimization of the weight vector means hill-climbing on this mountain to find maximum response. The sequential simplex algorithm is one of several strategies for searching a multi-dimensional variable set to find an optimum value for a response function [189, 243, 435].

The principle of the simplex algorithm will be described for a 2-dimensional optimization problem. The wanted weight vector $\bar{w}$ is

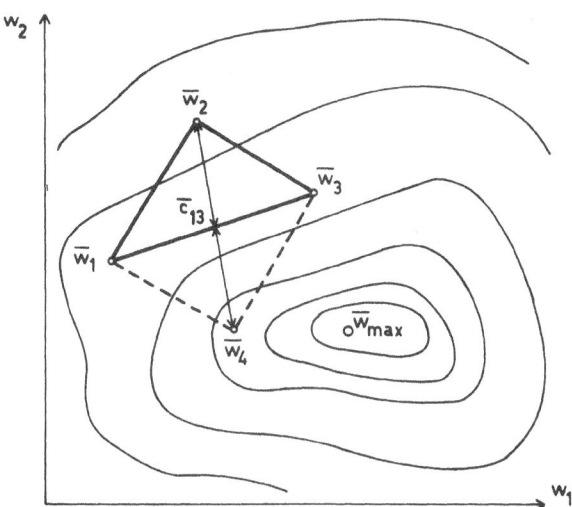

FIGURE 23. Simplex optimization. The response surface for 2-dimensional weight vectors $\bar{w}(w_1, w_2)$ with a maximum $\bar{w}_{max}$ is drawn as a relief map. The simplex algorithm starts with 3 arbitrarily selected weight vectors $\bar{w}_1, \bar{w}_2, \bar{w}_3$ which define a "simplex" (a triangle in this example). Movement the simplex away from the least desirable response leads to the maximum.

defined by 2 components w_1 and w_2. The performance (response) of $\bar{w}$ as a function of w_1 and w_2 is drawn in Figure 23 as a relief map. The simplex algorithm starts with 3 (= d+1) arbitrary selected weight vectors $\bar{w}_1, \bar{w}_2, \bar{w}_3$ which are the vertices of a triangle. A geometric figure with d+1 vertices in a d-dimensional space is called a <u>simplex</u>. The initial simplex is defined by $\bar{w}_1, \bar{w}_2$, and $\bar{w}_3$.

The maximum value of the response function is approached by movement of the simplex away from the least desirable response.

In our example $\bar{w}_2$ has the lowest response. To move the simplex against the maximum, $\bar{w}_2$ is reflected through the centroid $\bar{c}_{13}$ of the other d vertices and gives a new vertex $\bar{w}_4$. $\bar{w}_1, \bar{w}_3$ and $\bar{w}_4$ now span a new simplex which lies nearer the maximum than the previous simplex. This process is continued until no improvement is possible. An appropriate reduction of the simplex size and consideration of some precautions lead to the maximum of the response surface.

An assumption for this hill-climbing method is the existence of a continuous response surface with a unique extremum in the region of

search. In practical applications with no explicit knowledge of the
response function the algorithm is often stranded only at a local maximum or a plateau.

The computational effort for numerous dimensions – say $d \geq 100$ –
is considerable: d+1 different weight vectors must be initialized and
at least the same number of weight vectors must be stored; each new
weight vector requires classification of the whole training set. In the
next Chapters some aspects of the simplex algorithm are discussed in
more detail.

2.4.2. Starting the Simplex

The initial set of vertices should locate a region in the weight
vector space where the optimum of the response function is likely to
occur. The patterns of the training set are used to estimate this region.
One approach is to compute the centers of gravity of both classes; the
bisecting hyperplane (symmetry plane) defines a weight vector (Chapter
2.1.4) which can be employed as one of the initial vertices [243]. Also
the learning machine may be used to generate the first initial vertex.

TABLE 3. Generation of the initial simplex. The first initial weight vector $\bar{w}_1(w_1, w_2, \ldots w_d)$ is computed by the class means or by a learning
machine. Component w_d stems from the augmented pattern vector. The other
weight vectors $\bar{w}_2, \bar{w}_3, \ldots \bar{w}_{d+1}$ are generated from $\bar{w}_1$ by adding a spanning constant A to each of the components.

weight vector	weight vector components (coordinates)					
	1	2	3		d−1	d
$\bar{w}_1$	w_1	w_2	w_3	$\cdots$	w_{d-1}	w_d
$\bar{w}_2$	w_1+A	w_2	w_3	$\cdots$	w_{d-1}	w_d
$\bar{w}_3$	w_1	w_2+A	w_3	$\cdots$	w_{d-1}	w_d
$\cdots$						
$\bar{w}_{d+1}$	w_1	w_2	w_3	$\cdots$	$w_{d-1}+A$	w_d

The other initial weight vectors must be chosen to span the weight
space. One approach is to alter the first initial weight vector syste-
matically by adding a "spanning constant" A to each of the components
according to Table 3 [27, 242, 243].

More sophisticated methods for the initialization of the starting
simplex use for all features different spanning constants, depending on
the standard variations of the features [242].

2.4.3. Response Function

An obvious and easy to calculate optimization criterion is the
recognition rate. Because only integer numbers of patterns are classified, the response surface is discontinuous and consists of a series
of plateaus. It is possible for the simplex to become stranded on such
a plateau. Another problem arises if a weight vector with 100 % recognition is found which does not necessarily occupy the optimum position
between the clusters. Therefore, it is necessary to use a second optimization criterion to smooth the surface.

If two or more vertices of the simplex have the same value for the
first response function, the value of the second response function is
used to find the least desirable vertex. Ritter, Lowry, Isenhour, and
Wilkins [242] proposed the minimization of the

perceptron function :
$$\sum_j |\bar{w}.\bar{x}| \quad \text{---->} \quad \text{minimum} \tag{63}$$

$$\text{or} \quad \sum_j \tanh |\bar{w}.\bar{x}| \quad \text{---->} \quad \text{minimum} \tag{64}$$

or the

squared error function :
$$\sum_j (\bar{w}.\bar{x})^2 \quad \text{---->} \quad \text{minimum} \tag{65}$$

Summation is always carried out over all misclassified patterns [376].
Another strategy is to minimize the distance between the decision
boundary and the nearest misclassified pattern. Further response functions are discussed in [242].

2.4.4. Moving the Simplex

The response is computed for all d+1 vertices of the simplex. The vertex $\bar{w}_w$ with the worst response is reflected through the centroid $\bar{w}_c$ of the other vertices. The new vertex $\bar{w}_r$ is given by (Figure 24)

$$\bar{w}_r = \bar{w}_c + (\bar{w}_c - \bar{w}_w) \tag{66}$$

$$\bar{w}_c = \frac{1}{d} \sum_{\substack{j=1 \\ j \neq w}}^{d} \bar{w}_j \tag{67}$$

Care must be taken that the same vertex is not considered as inferior in consecutive iterations.

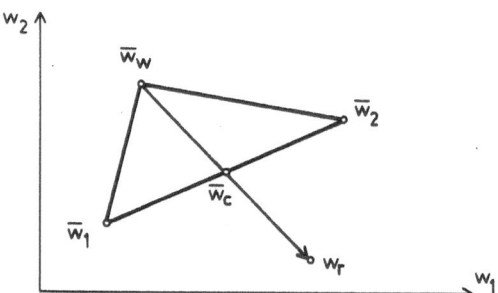

FIGURE 24. Reflexion of the vertex $\bar{w}_w$ with the worst response through the centroid $\bar{w}_c$ of the other vertices $\bar{w}_1$, $\bar{w}_2$ gives the new vertex $\bar{w}_r$.

In the modified simplex method [425] further steps are necessary to compute an optimum new vertex and to find the maximum response with maximum speed and efficiency :

1. If the reflected vertex $\bar{w}_r$ is better than the best vertex $\bar{w}_b$ of the simplex, correct movement is indicated and the simplex is further expanded to $\bar{w}_e$ (Figure 25).

$$\bar{w}_e = \bar{w}_c + 2 (\bar{w}_c - \bar{w}_w) \tag{68}$$

If $\bar{w}_e$ has a better response than $\bar{w}_r$, then $\bar{w}_e$ is used as the new vertex; otherwise, $\bar{w}_r$ is the new vertex.

2. If the reflected vertex $\bar{w}_r$ is less desirable than the worst vertex $\bar{w}_w$ in the simplex, then $\bar{w}_t$ is taken as the new vertex (Figure 25).

$$\bar{w}_t = \bar{w}_c - (\bar{w}_c - \bar{w}_w) / 2 \qquad\qquad (69)$$

3. If the reflected vertex $\bar{w}_r$ is better than $\bar{w}_w$ but worse than the second best vertex of the simplex, a vertex $\bar{w}_z$ is used as the new vertex (Figure 25).

$$\bar{w}_z = \bar{w}_c + (\bar{w}_c - \bar{w}_w) / 2 \qquad\qquad (70)$$

4. If the reflected vertex $\bar{w}_r$ is better than the second best vertex but worse than the best vertex $\bar{w}_b$, then $\bar{w}_r$ is taken as the new vertex.

The modified simplex algorithm produces nearly optimum weight vectors even when linearly inseparable data sets are used for training but has the disadvantage of sometimes prohibitively large computer times [165].

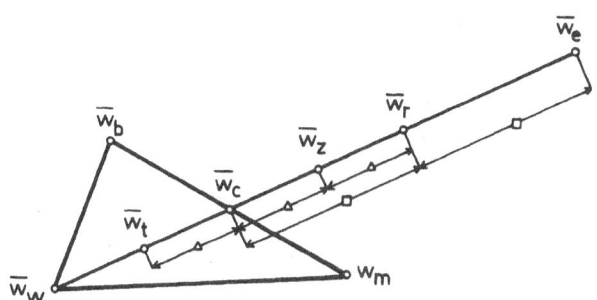

FIGURE 25. Modified simplex method. Instead of the worst vertex $\bar{w}_w$ one of the vertices $\bar{w}_t$, $\bar{w}_r$, $\bar{w}_e$ is used as a new vertex.

A further variation of the simplex method – the super-modified simplex method – was developed by Denton et. al. [432] and used in chemical pattern recognition by Kaberline and Wilkins [137]. For each simplex the responses at the worst vertex $\bar{w}_w$, at the centroid $\bar{w}_c$ and at the point $\bar{w}_r$ are calculated (Figures 25 and 26). A second-order polynomial curve is fitted to these three points. Furthermore, the curve is

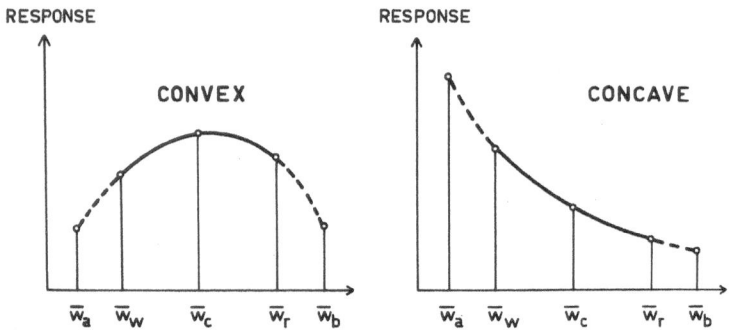

FIGURE 26. Super-modified simplex. Possible response curves. The curves are extrapolated beyond $\bar{w}_w$ and $\bar{w}_r$ by a fixed fraction F of the $\bar{w}_w$ to $\bar{w}_r$ distance [137].

extrapolated beyond $\bar{w}_w$ and $\bar{w}_r$ by a fixed fraction F (e.g. F=0.5) of the $\bar{w}_w$ to $\bar{w}_r$ distance. Two distinct curve types may result (Figure 26).

For the convex case a maximum exists somewhere. The maximum is computed from the first derivative of the curve. Three distinct situations may arise :

1. Location of the maximum within the region $\bar{w}_a$ to $\bar{w}_b$. For this case, the new vertex is located at the maximum point of the curve.

2. Location of the maximum is outside the region $\bar{w}_a$ to $\bar{w}_b$. Here, the maximum point within the region $\bar{w}_a$ to $\bar{w}_b$ is used for the new vertex.

3. If the maximum lies very close to the centroid $\bar{w}_c$, the dimensionality of the new simplex would be reduced. A small "safety factor" (e.g. 5 % of the distance $\bar{w}_w$ to $\bar{w}_r$) is necessary to locate the new vertex a distance away from the centroid [137].

For the concave case, no maximum occurs within the region $\bar{w}_a$ to $\bar{w}_b$. In this case, the boundary with the highest response is used for the new vertex.

The ability of the super-modified simplex algorithm to locate the position of the new vertex more precisely than is possible by the modified simplex method means that the super-modified simplex requires fewer iterations to converge and is therefore somewhat faster.

A high speed algorithm for simplex optimization calculations in pattern recognition, similar to the super-modified simplex, has been described by Brissey et. al. [24].

2.4.5. Halting the Simplex

If the data of the training set are linearly separable, the simplex optimization should terminate at 100 % recognition rate. For data appearing linearly inseparable, a procedure was proposed by Ritter, Lowry, Wilkins, and Isenhour [242, 243] to leave poor local maxima of the response function. If the response change is less than 1 % over ten consecutive iterations, a new simplex is re-expanded about the optimum weight vector (the new first initial weight vector $\bar{w}_1$ in Table 3 is the best vector found in the last iteration). The simplex algorithm is then applied again.

The convergence time of the super-modified simplex was found experimentally to be proportional to the training set size and proportional to the square of the number of dimensions [242].

2.4.6. Chemical Applications

The simplex method was originally proposed by Spendley, Hext, and Himsworth [435] in 1962 and modified by Nelder and Mead [425] in 1965. The simplex optimization has been successfully applied to several areas of analytical chemistry such as experimental optimization [189, 414], data reduction [426] and instrument control [415].

Applications of the simplex method in pattern recognition were examined by Isenhour, Wilkins et. al. for the interpretation of mass spectra [137, 165, 242, 243], and carbon-13 nuclear magnetic resonance spectra [26, 27].

The simplex algorithm generated better linear, binary classifiers than the learning machine or a distance measurement approach. The advantages of the simplex method are stressed for linearly inseparable data and for a small number of features. A disadvantage is the large computational effort which is necessary to obtain a classifier (up to more than one hour CPU-time using a modern large computer) [137].

The application of pattern recognition to the recognition of chemical substructures on the basis of mass spectra did not reveal considerable advantages of the super-modified simplex over the modified simplex [137]. Extensive examinations of the behaviour of the simplex algorithm in pattern recognition have been made with artificial data [242].

2.5. Piecewise-Linear Classifiers

In certain classification problems, a linear separation of two
classes by only one decision plane is impossible. Figure 27 shows a
two-modal class (+) consisting of two distinct clusters (subclasses).
Evidently, this class should be represented by two prototypes ($\bar{w}_a$, $\bar{w}_b$)
and a minimum distance classifier would be successful. In this way, the
pattern space is partitioned by several decision planes (piecewise-
linear separation). Classification of an unknown pattern requires the
calculation of the scalar products with all weight vectors (prototype vectors). The unknown is assigned to the class with the largest
scalar product (Chapter 2.1.5.). In the same way, a multicategory
classification is possible [89, 396].

For a d-dimensional classification problem is it not possible to
define a priori an optimum number of prototypes. Chemical knowledge or
intuition about existing subclasses is helpful for the generation of
initial weight vectors. Centres of gravity of these subclasses may be
used as an initial set of weight vectors. An iterative process with an
error correction similar to that of the learning machine (Chapter 2.2)
is then used to train the weight vectors [89, 396].

Suppose a pattern $\bar{x}$, actually belonging to subclass m, is presented to all prototypes and subclass l yields the largest scalar product.
The weight vector $\bar{w}_m$ which should give the maximum scalar product is
corrected by equation (71).

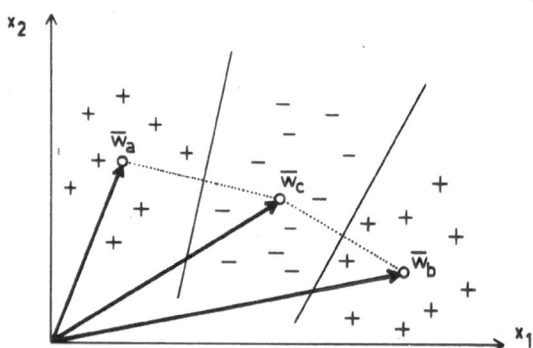

FIGURE 27. Piecewise-linear classification. The two-modal class (+) is
represented by two prototypes $\bar{w}_a$ and $\bar{w}_b$.

$$\bar{w}_m^{\,\prime} \quad = \quad \bar{w}_m \quad + \quad c.\bar{x} \tag{71}$$

The weight vector $\bar{w}_l$ which erronously gave the largest scalar product is corrected by equation (72).

$$\bar{w}_l^{\,\prime} \quad = \quad \bar{w}_l \quad - \quad c.\bar{x} \tag{72}$$

Factor c is some positive correction increment (Chapter 2.2.3).

If a weight vector continuously oscillates during the training, then an additional weight vector is necessary to obtain piecewise-linear separability of the training set. Frew, Wangen, and Isenhour [89] proposed an algorithm to recognize such oscillations and to generate appropriate additional weight vectors until the training converges.

Application of a multicategory piecewise-linear classifier to the interpretation of mass spectra was of "varying success" [88, 89, 117]. Recognition of a C=C double bond in a linearly inseparable data set of mass spectra required 3 to 5 weight vectors to obtain 78 to 87 % predictive ability (75 % was obtained with a single weight vector for the same data set). The lack of a satisfactory theory of piecewise-linear classifiers and rather high computational expenses have prevented up to now broader applications of this classification method.

2.6. Implementation of Binary Classifiers

2.6.1. Discrete or Continuous Response

Each classification contains several steps as shown in Figure 28. The components of a pattern vector $\bar{x}$ are used by a discrimination function f to compute a discriminant $s = f(\bar{x})$. The scalar s can be employed as a continuous response of the classifier. The magnitude (and sign) of s can therefore be used either to predict a continuous property of the unknown or to estimate the confidence of an assignment to a discrete class (Figure 29).

The continuous discriminant s is usually compared with some constant value in a "threshold logical unit" (TLU). The output of the TLU is used

as a discrete response of the classifier. The optimum threshold s_0 for
a binary classification is indicated in Figure 29. A classifier with a
continuous response gives a more detailed result than one with a discrete
response but requires a larger training set and more computational
efforts.

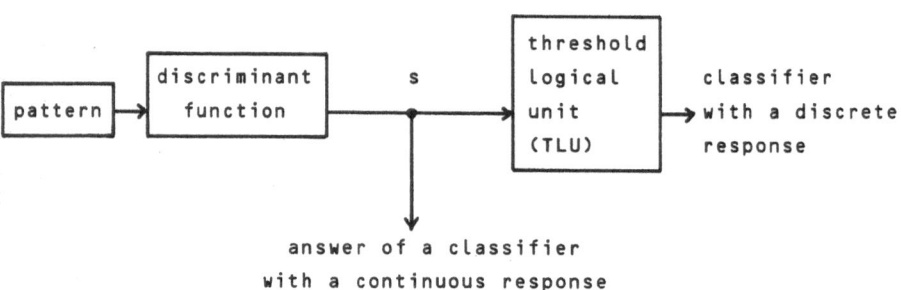

FIGURE 28. Classifier with continuous or discrete response.

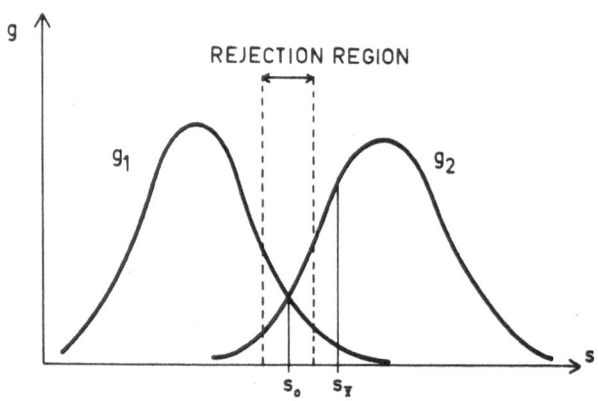

FIGURE 29. Probability density functions for a binary classifier with
a continuous response s; g_1 and g_2 are the probability densities for
class 1 and class 2, resp. as a function of response s. If an unknown
pattern $\bar{x}$ yields the response $s_{\bar{x}}$ then $g_1(\bar{x})$ and $g_2(\bar{x})$ give the probabilities of class 1 and 2. The region of similar values for g_1 and g_2 may
be used for a rejection of the classification. s_0 should be employed as
an optimum threshold for a TLU.

2.6.2. Classification by a Committee of Classifiers

A "committee machine" uses several different classifiers for the
same classification problem in a parallel manner. The classifiers may
stem e.g. from different training processes with the same training set.
The results (scalar products) are summarized to give the final classification. Threshold logical units are usually applied and the majority of
votes determines the class membership. Correct classification for all
members of the training set may be obtained by a committee machine even
if the data set is not linearly separable. Unfortunately, no exhaustive
theory about the training of several parallel classifiers exists.

A committee machine is a special case of a "layered machine".
A layered machine consists of several levels of classifiers (TLU's).
The outputs of one level are the inputs for the next level; input for
the first level is the pattern to be classified; output of the last
level (containing only one TLU) is the final response of the classification machine [396].

2.6.3. Multicategory Classification

A discrimination between several mutually excluding classes requires
the application of a multicategory classifier or that of a series of binary classifications. Some methods which have been described for binary
classifiers can be altered for multicategory problems. Suitable methods
for determining a multicategory response are classification by distance
measurements to centres of gravity, linear regression and piecewise-
-linear classifiers. A multicategory classification problem is e.g. the
prediction of the number of atoms of a certain element per molecule
using a spectral pattern. If the number to be predicted lies e.g. between 0 and 5, then a discrimination of 6 classes is necessary.

Three different strategies are possible if a series of binary decisions is used to solve a multicategory problem [85]:

a. Parallel Classifications
In this method each binary classifier is trained to classify patterns
into one of two classes separated by a cutoff point. Class 1 means

greater than the cutoff. Each classifier is trained separately with the whole training set. A parallel arrangement of all these classifiers is used for the classification of unknowns (Table 4). The number of atoms is given by the cutoff of the first classifier yielding an assignment to class 2. Discontinuities in the classification responses indicate erroneous decisions.

TABLE 4. Parallel arrangement of binary classifiers for the prediction of the number of atoms of a certain element in the molecule.

binary classifier	number of atoms in class 1	class 2	assigned class for a triatomic molecule
$\bar{w}_a$	> 0	0	1
$\bar{w}_b$	> 1	$\leq$ 1	1
$\bar{w}_c$	> 2	$\leq$ 2	1
$\bar{w}_d$	> 3	$\leq$ 3	2 (break !)
$\bar{w}_e$	> 4	$\leq$ 4	2

b. Branching Tree Classifications
In this method, the binary classifiers are arranged in a branching network. Each classifier is trained to dichotomize a set of pattern vectors according to the scheme in Figure 30. For the training, only those patterns pertinent to a distinct branch point should be used. The method suffers from the accumulation of errors for successive decisions.

c. Binary Code Classification
The number of the class (>2) is coded as a binary number. For each binary digit, a binary classifier is trained that predicts "zero" or "one". This classification method requires the smallest number of binary classifiers [85, 178]. For example, up to 8 classes can be discriminated by 3 binary classifiers (Table 5). The accuracy of the method can be improved by introducing additional binary digits (additional binary classifiers) to form an error-correcting code analogous to a "parity bit".
However, for some classifiers (binary digits), a physically meaningful

separation into two classes seems very doubtful. Thus, the classifier $\bar{w}_b$ in Table 5 has to discriminate between class numbers 0, 1, 4, 5 (2nd binary digit is '0') and class numbers 2, 3, 6, 7 (2nd binary digit is '1'). Nevertheless, surprisingly good results have been reported for this method. The carbon number (4 to 10) of hydrocarbons was correctly predicted from low-resolution mass spectra in 94 % [85].

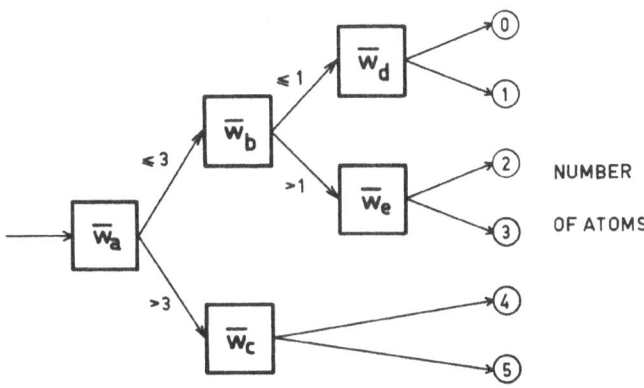

FIGURE 30. Branching tree arrangement of binary classifiers $\bar{w}$ for a multicategory problem.

TABLE 5. Multicategory classification by binary classifiers that recognize the digits of the binary encoded class number. A set of 3 classifiers $\bar{w}_a$, $\bar{w}_b$, $\bar{w}_c$ may discriminate 8 different classes.

response of the classifiers			class number
$\bar{w}_a$	$\bar{w}_b$	$\bar{w}_c$	
0	0	0	0
0	0	1	1
0	1	0	2
0	1	1	3
1	0	0	4
1	0	1	5
1	1	0	6
1	1	1	7

3. K - Nearest Neighbour Classification (KNN-Method)

3.1. Principle

The $\underline{K}$-$\underline{N}$earest $\underline{N}$eighbour (KNN) classification is a standard method
in pattern recognition and is especially marked out by its simplicity
[128, 148, 305, 370, 374, 379, 389].

Each pattern is characterized in the usual way (Chapter 1.2) by a
set of d components (features, measurements) and can be considered as a
point in a d-dimensional space. An additional component as described in
Chapter 1.3 is not necessary for the KNN-method. Classification of an
unknown pattern $\bar{x}$ is made by examination some pattern points with known
class membership which are closest to $\bar{x}$. In order to find the nearest
neighbours of the unknown it is necessary to compute the distances between $\bar{x}$ and all other pattern points of the available data set. The number of neighbours which are considered for classification is usually denoted by 'K'. If only one neighbour is used for the classification (K=1,
"1NN-method") the class membership of the first (nearest) neighbour
gives the class membership of the unknown. If more than one neighbour is
used a voting scheme or some other procedure is applied to determine the
class of the unknown.

The distance between two points $\bar{x}(x_1, x_2, \ldots x_d)$ and $\bar{z}(z_1, z_2, \ldots z_d)$ in the d-dimensional pattern space is usually defined by the
Euclidean distance D

$$D = \left[\sum_{i=1}^{d} (x_i - z_i)^2 \right]^{1/2} \tag{73}$$

but any other metric can be employed (Chapter 2.1.9).

The KNN-method does not require linearly separable clusters
(Figure 31). The KNN-method is a multiclass method. Membership to any
number of classes is determined by the same neighbour patterns!
No training is necessary because the classification procedure contains
all patterns of the data set. New patterns may be added to the data set
without difficulties. The main disadvantage of the original KNN-method
is the fact that no data compression is possible; all pattern vectors

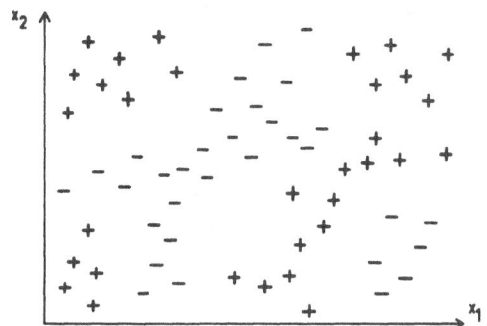

FIGURE 31. KNN-method. In this linearly inseparable data set, each point
- considered as unknown - is classified correctly by the first neighbour.

must be stored and many computations are necessary to find the nearest
neighbours. Therefore, the KNN-classification is especially suited for
small data sets with few dimensions. Because of the statistical foun-
dation of the maximum risk (Chapter 3.2) the KNN-method serves as a
standard method in pattern recognition for comparisons with more sophisticated classification procedures.

3.2. Maximum Risk of KNN-Classifications

If the d-dimensional probability densities of all classes are
known, an average minimal risk of classification is given by the
expected Bayes risk (Chapter 5.2.). No classification method - either
parametric or non-parametric - is possible with a smaller risk than the
expected Bayes risk. Cover and Hart [374] showed for a 2-class problem
that the maximum risk of the KNN-method (K=1) is twice the expected
Bayes risk. Therefore, any other classification method can, at most,
double the performance of the 1NN-method, i.e. half of the available
classification information in a very large data set is contained in
the nearest neighbour. The greater the number of samples in the training set, the closer the risk will approach the lower limit [152].

3.3. Characteristics and Variations of the KNN-Method

If the magnitudes of the features are not similar, a scaling should precede the classification [44, 242].

Besides the Euclidean distance other distance measurements (Chapter 2.1.8) have been used for chemical applications of the KNN-method. The normalized distance is independent of the number of dimensions [175]; the Hamming and Tanimoto distances are suitable for binary encoded patterns [353, 356, 357].

In most chemical applications of the KNN-method, only the first (nearest) neighbour (K=1) was used for classification. For K>1 a simple voting ("one neighbour one vote") may be applied. The contributions of the neighbours to the voting can also be weighted by the distances (or the squared distances) between the unknown and the neighbours.

$$V_{total} \; = \; \sum_{j=1}^{K} \frac{V_j}{D_j} \qquad or \qquad V_{total} \; = \; \sum_{j=1}^{K} \frac{V_j}{D_j^2} \qquad (74)$$

V_j : +1 or -1 depending on whether neighbour j belongs to class 1 or 2
D_j : any distance measurement for the distance between unknown pattern and neighbour j
K : number of neighbours

The unknown is grouped into class 1 if the voting result V_{total} is positive and to class 2 for a negative result [44, 133, 135].

For an even number of neighbours, the voting may end undecided and the classification of the unknown pattern has to be rejected. Rejection of difficult to classify patterns usually increases the success rate for the remaining actually classified patterns [306].

Whether more than one neighbour should be used for classification or not obviously depends on the classification problem - contradictory results have been reported. A class with only a few members in the data set is certainly disadvantageously affected by a large number of neigh- bours [135, 201].

An increase in the number of neighbours permits to estimate the probabilities of all classes at the point which is defined by the unknown pattern. With a large number of neighbours the KNN-method becomes a modified Bayes classification (Chapter 5) [200].

An alternative technique to conventional KNN-classification was proposed by Pichler and Perone [228]. A feature selection procedure based on the KNN-technique gives first a set of m best features for a certain classification problem (Chapter 10). All m features are used individually for 1-dimensional KNN-classifications of unknowns. Voting is accomplished by the responses of the features ("consensus vote technique").

3.4. Classification with Potential Functions

A powerful yet laborious classification method uses potential functions to decide on the class membership of a point $\bar{x}$ in the pattern space. At each pattern point of the data set an "electric charge" is located. Each pattern is therefore surrounded by its own potential field. At a distance D from the pattern point the potential Z(D) may be given by equation (75) [369] or equation (76) [76].

$$Z(D) \quad = \quad \frac{1}{1 + qD^2} \tag{75}$$

or

$$Z(D) \quad = \quad e^{-\frac{D^2}{q}} \tag{76}$$

q describes the form of the potential function.

Superposition of the potentials from all patterns belonging to the same class gives the overall potential of that class in all parts of the pattern space. This potential is normalized to the number of sources and may be used as an approximation of the probability density for that class [389: p.98].

An unknown pattern $\bar{x}$ is grouped into the class exerting the strongest field at point $\bar{x}$. For a binary classification problem all members of class 1 are given a positive unit charge and all members of class 2 a negative unit charge.

$$z(\bar{x}) \quad = \quad \frac{1}{n_1} \sum_{j=1}^{n_1} z_1(\bar{x}_{j1}) \quad - \quad \frac{1}{n_2} \sum_{j=1}^{n_2} z_2(\bar{x}_{j2}) \tag{77}$$

$z_m(\bar{x}_j)$: potential of pattern $\bar{x}_j$ at the position of the unknown $\bar{x}$,
   z_1 for patterns of class 1, z_2 for patterns of class 2
$\bar{x}_{j1}$: pattern j of class 1
$\bar{x}_{j2}$: pattern j of class 2
n_1 : number of patterns in class 1
n_2 : number of patterns in class 2
$z(\bar{x})$: superposition potential at the position of the unknown $\bar{x}$,
   positive: classification into class 1
   negative: classification into class 2

The decision boundary is given by all points with zero potential;
the boundary is usually non-linear and non-continuous (Figure 32).

In principle, the form of the potential function and other controlling parameters can be varied as a function of the class membership, region of pattern space and the probability of the presence
of different classes to achieve optimum classification. However, for
practical applications, this optimization seem to be impracticable [128].

The overall potential of a more frequently occurring class may completely superimpose the overall potential of classes containing a small
number of members. To overcome this difficulty a training process can be
used to enlarge stepwise the electric charge of patterns from the less
frequent class (Figure 33). This training process does not necessarily
converge [369].

In a simpler and faster version of the classification by potential
functions, only a number of neighbours nearest to the unknown are used
for the calculation of the overall potentials.

The number of patterns in the training set can be reduced to those
patterns which are necessary to classify all patterns correctly [76].

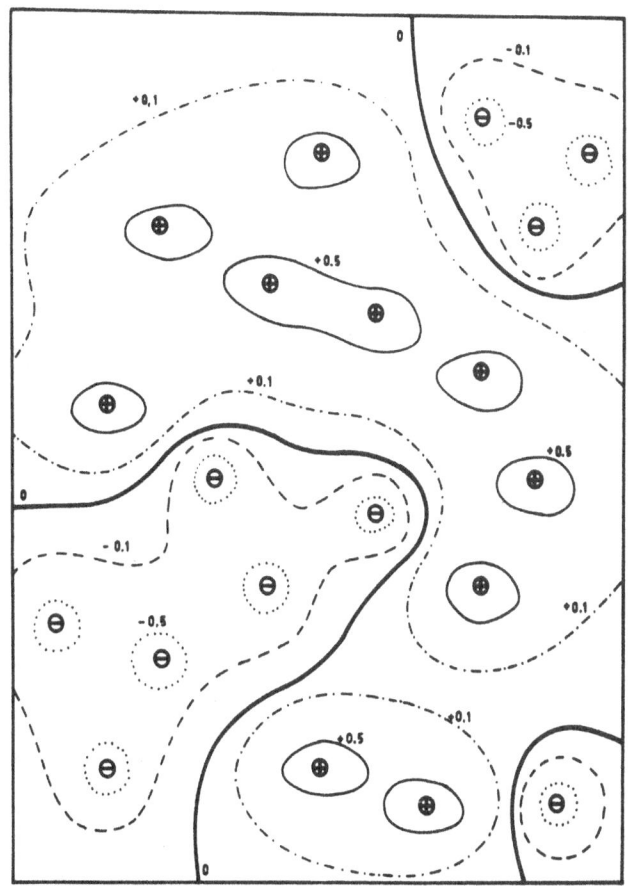

FIGURE 32. Classification with potential functions. To each member of class 1 and 2 was assigned a unit charge (positive or negative, resp.). Lines of equal superposition potential Z were calculated by equation (75) for $q = 0.1$. The decision boundary between the two classes is equal to the line $Z = 0$. The absolute magnitude of Z can be used as a confidence measure.

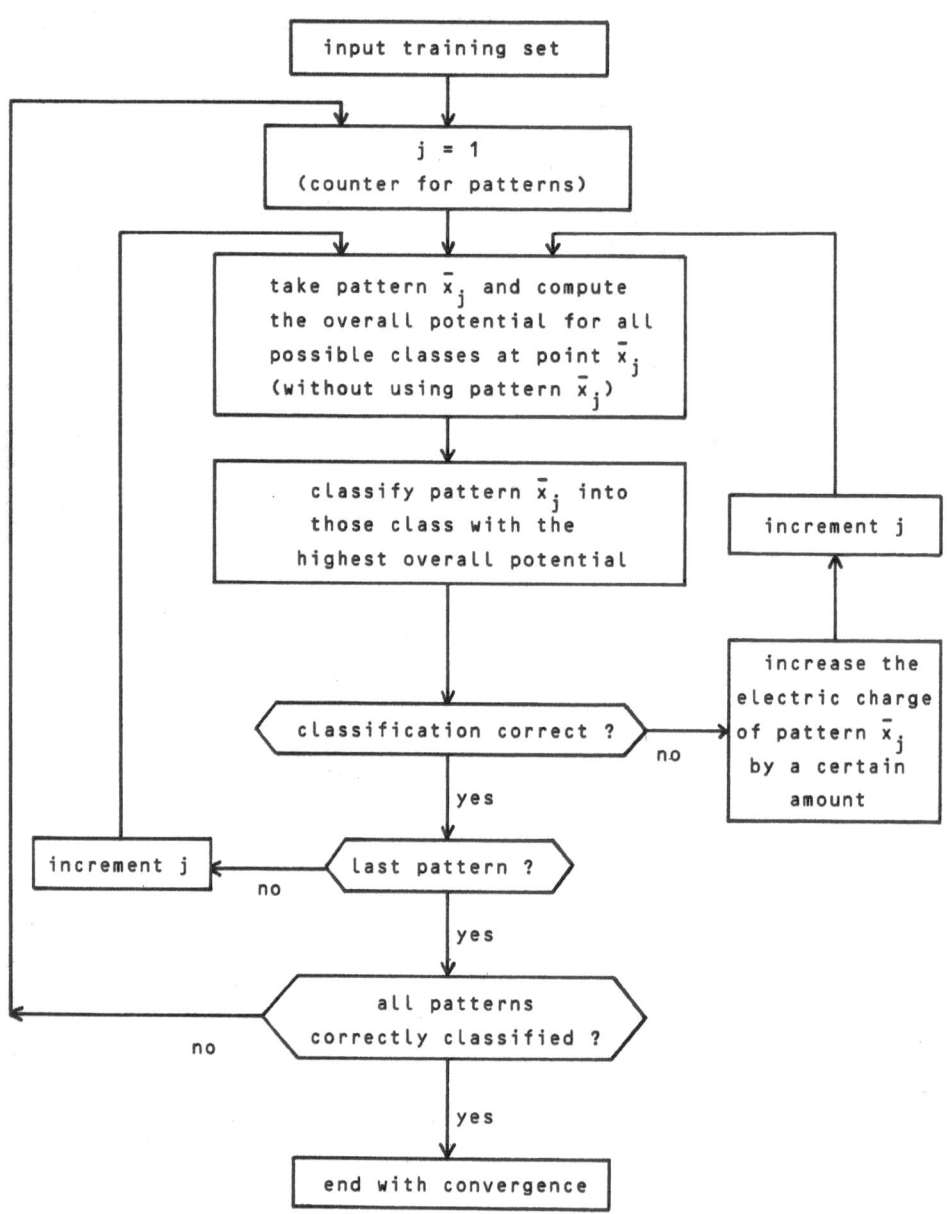

FIGURE 33. Classification by potential functions. Adjustment of "electric charges" in the data set [369].

3.5. KNN-Classification with a Condensed Data Set

A disadvantage of the KNN-method is the large amount of time which
is required to classify unknown patterns with a large data set. Because
all patterns of the data set must be examined to classify each unknown,
the computational requirements may make applications prohibitively expensive. Classification time can be significantly decreased if the training set can be reduced to a smaller number of patterns which lie near
the decision boundary. Several strategies have been proposed to find an
optimum subset with a minimum number of patterns that correctly classifies all patterns of the original data set.

The edited KNN-method eliminates patterns that are incorrectly
classified by use of the remainder of the data [405, 406].

Ritter et. al. [246] described a selective KNN-method to approximate the decision boundary by an optimum subset of patterns.

The condensed nearest neighbour rule generates a consistent subset of patterns (Figure 34). This subset, when used as a stored reference
set for the KNN-rule, correctly classifies all of the remaining patterns.
In general, the algorithm will not find a minimum consistent subset but
may end with a subset equal to the original set [297, 380, 383].

A similar algorithm - the reduced nearest neighbour rule - was
described by Gates [382].

3.6. Chemical Applications

A KNN-classification is almost identical with the interpretation of
spectra by a library search. In library search an unknown spectrum
(pattern) is compared with all spectra of known compounds collected
in a spectral library. A similarity criterion or a dissimilarity
criterion (equivalent to a distance measurement) between two spectra
must be defined. To find the most similar spectra in the library, this
criterion must be calculated for each library spectrum.

However, interpretation of the results is different in library
search and in KNN-classifications. Pure library search looks for an
identification of the unknown and is often useless if the library does
not contain the unknown. KNN-classification restricts the answer to a

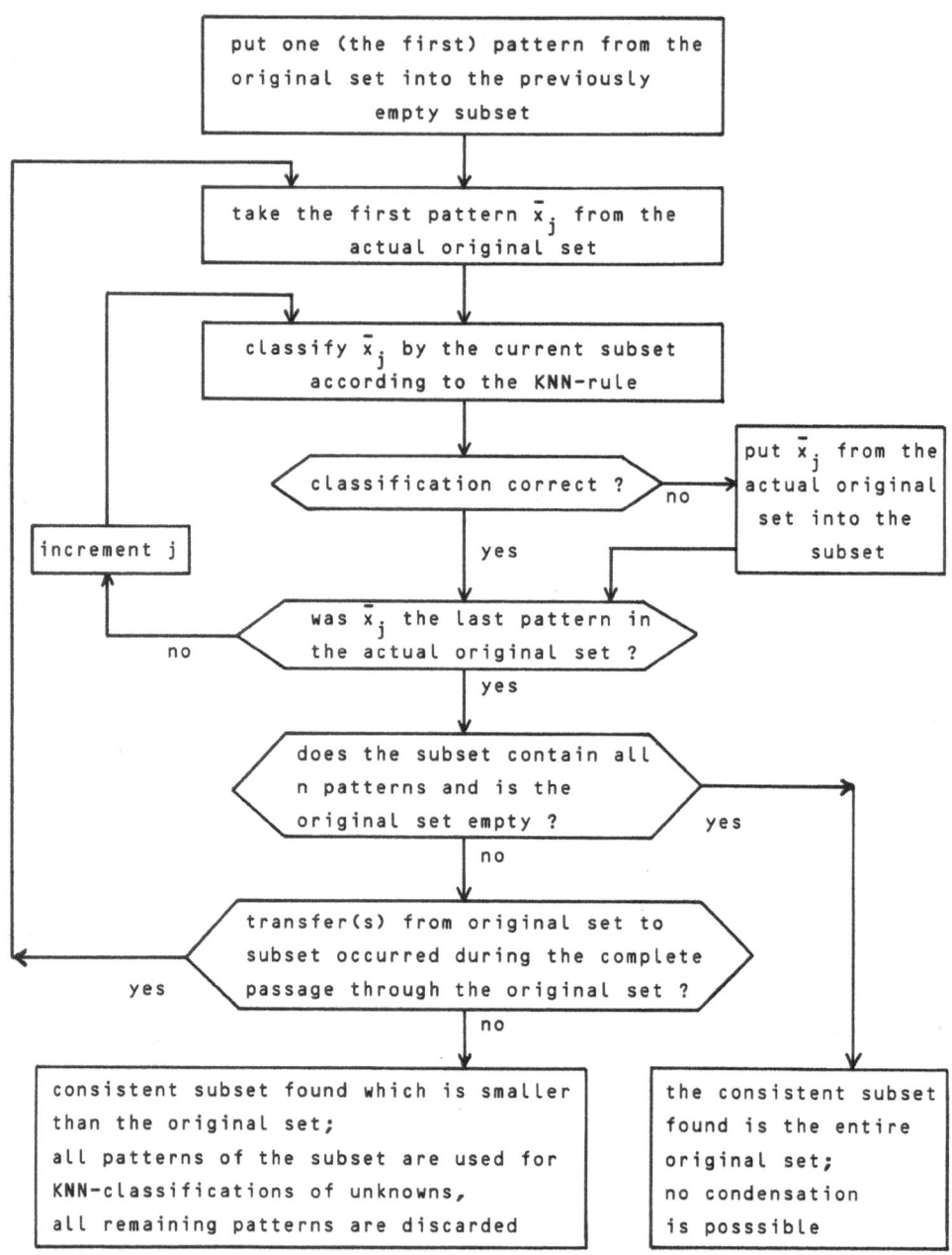

FIGURE 34. Condensed nearest neighbour method to reduce the original
data set of n patterns to a consistent subset that correctly
classifies all patterns of the original set [383].

statement about the membership to a certain class (or to several classes). The quality of this answer can be estimated by the leave-one-out
method (Chapter 1.4.). Each pattern of the data set is treated as unknown
and classified by all other patterns. If the spectral library contains
only one spectrum for each compound, all classifications in the test are
made with "strange spectra". Therefore, this test simulates practical
conditions in which unknowns have to be classified by a data set which
does not contain the unknowns.

The KNN-rule has been applicated to a great number of chemical
classification problems. The interpretation of mass spectra [135, 251,
305, 306], nuclear magnetic resonance spectra [152], infrared spectra
[76, 353], and polarographic data [228, 297] has been accomplished successfully by the KNN-method. The performance of the KNN-algorithm was
generally better or at least as good as for other pattern recognition
methods.

The KNN-method has been used to analyze 2-dimensional projections
of d-dimensional clusters in order to evaluate the method of projection
[248, 299].

The KNN-method is the method of choice if the cluster structure is
complex and a linear classifier fails. Because of the large computational
requirements necessary for KNN-classifications the method is not suitable
for a large number of unknowns or a large data set of known patterns.
Good performance as well as simplicity and variability of the KNN-method
should be taken into account if the computer time for a classification
is less important than a maximum of success.

The best descripition of the data is the data themselves !

4. Classification by Adaptive Networks

4.1. Perceptron

Starting from some crude ideas about the structure of the human brain and the human eye an interesting pattern recognition machine - called perceptron - was developed by Rosenblatt [399]. Figure 35 shows the structure of a perceptron for binary classifications. The "machine" needs not be electronically wired but can be simulated with a computer program.

The input device of the machine consists of a series of receptors. Each receptor corresponds to one component x_i of a binary encoded pattern $\bar{x}$ (x_i has the value 0 or 1). The second layer of the machine consists of a series of association units. Each association unit has several inputs but only one output. The inputs of the association units

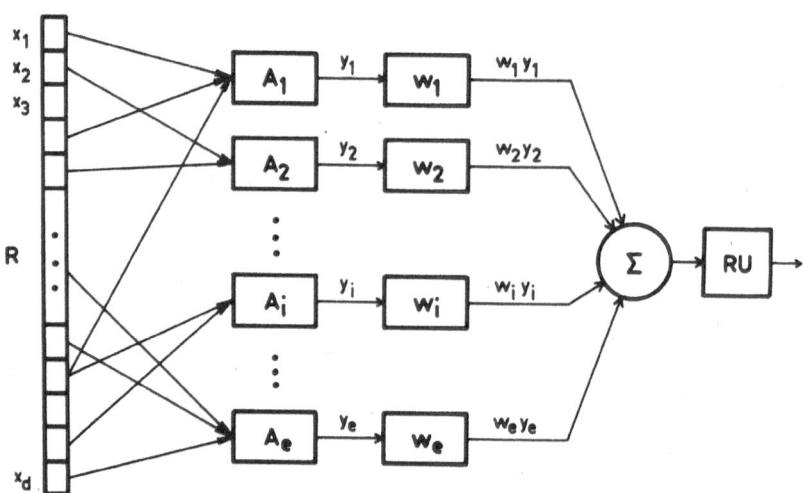

FIGURE 35. Perceptron. R: receptors (with binary encoded outputs $x_1 \ldots x_d$, equivalent to a binary encoded pattern vector), A: association units (with binary outputs $y_1 \ldots y_e$), $w_1 \ldots w_e$: adjustable weights for the outputs of the association units, RU: response unit.

(not necessarily all) are arbitrarily connected with receptor outputs.
Furthermore, each input of the association units is given randomly a
positive or negative sign. The connections remain the same for one machine (experiment). An association unit sums the input signals (-1, 0,
or +1) and compares the result with a threshold t (t may be the same
and constant for all association units). If the sum is larger or equal to
t, the association unit is called "excited" and a logical '1' is generated at the output y. If the sum is smaller than t the output y is '0'.
Each output y_i of the association units is multiplied by an adjustable
weight w_i and all products are summed (equivalent to the calculation of
a scalar product). The weights w_i may be negative, zero or positive.

A response unit compares the result (scalar product) with a threshold and produces the final classification result, e.g. a '1' for
positive scalar products (class 1) and a '0' for negative scalar products
(class 2).

Weights and thresholds are adjusted during the training phase.
Patterns with known class membership are presented to the machine.
Depending on the answer of the machine the weights are changed or not.
After a certain number of training patterns the perceptron should be able
to classify correctly all or almost all patterns. A negative feedback
algorithm was successfully applied to the recognition of printed characters:
- The weights are not changed if the classification was correct.
- If a misclassification occurs only those weights are corrected which
  correspond to excited association units.
- Correction means that the weights for all association units which
  were excited by the misclassified pattern are increased if the scalar
  product was too small; the weights are decreased if the scalar product
  was too large.

A geometrical interpretation of the perceptron shows that the
pattern space is divided by the association units into several poly-
hedrons. The partitioning is random because the receptors are randomly
connected to the association units. The perceptron coordinates each
polyhedron either to class 1 or to class 2.

The perceptron consists of two main parts. In the first part a
new pattern vector $\bar{y}$ is calculated from each original pattern vector $\bar{x}$.
Each component y_i is a linear combination of several randomly selected
components from the original pattern. The second part is an application
of a binary, linear classifier and all methods from Chapter 2 may be
used.

A random combination of features is a reasonable approach if no
physically meaningful combinations of features are obvious for a given
classification problem.

4.2. Adaptive Digital Learning Network

This classification procedure was originally developed by Bledsoe
and Browning [372] and first applied to chemical problems by Stonham,
Aleksander, et.al. [284]. The learning network has some similarities
to the perceptron, especially in the random combination of features
[366, 367].

The patterns used with this method are binary encoded patterns $\bar{x}$.
Groups (n-tuples) of n (usually 3 or 4) pattern components are randomly
chosen and associated with a "memory element". A memory element has
2^n addressable 1-bit-storage locations. The configuration of the bit
pattern of the n-tuple is used to address one of the 2^n locations.
Example: For n = 4 a four-bit binary pattern is interpreted as one of
the decimal numbers 0 to 15. In the training stage a '1' is written into
the location addressed by the n-tuple.

A "digital learning network" consists of a group of memory elements.
For example, if patterns with 100 features are to be classified and
n = 4, then a set of 25 16-bit-memory elements are necessary (with no
feature being sampled twice). All elements are initialized to '0' and
connected randomly to the pattern components.

To train a digital learning network for a particular class, only
patterns belonging to that class are presented to the network. Each
n-tuple of features selects a storage location. If the selected storage
already contains a '1' nothing happens, otherwise a '1' is written
(an alternative training method is described below).

When recognition of a pattern is desired, the pattern is presented
to the trained network just as for training. All contents of the storage locations being addressed by the n-tuples of the pattern components
are summed. If a pattern of the training set is classified by this procedure, all storage locations addressed will have been set to '1' during
the previous training; the sum will be equal to the number of memory
elements. This is the maximum response that a pattern can achieve. For
a binary classification of an unknown the sum is compared to a threshold;

if the sum is greater than the threshold the pattern is placed into that class for which the network was trained. For a multicategory classification separate networks have to be trained for each class. An unknown is presented to each trained network and the maximum response determines the class membership.

The digital learning network may be implemented with hardware or simulated with a computer program. Several characteristics of the method have been investigated by Stonham et.al. [282 - 285].

The method of binary template matching is equivalent to a learning network with n = 1. A binary template of a class is the superposition (logical "and"-function) of all binary encoded patterns of that class. An unknown is compared with a template by counting the number of identical bit positions.

If the training set is too large an "overtraining" ("over-generalization") may occur because almost all possible bit combinations occur at least once and therefore almost all storage locations are filled with '1's. To overcome this difficulty Stonham and Shaw [285] proposed the generation of an optimum training set (Figure 36).

Soltzberg et.al. [280] modified the digital learning network classification by the use of "integer memory arrays" instead of "binary memory arrays". During the training the content of a particular location of the memory is incremented each time that location is addressed by an n-tuple. Since the values of the memory locations depend on the number of patterns used in the training, a normalization of the memory contents is necessary. This was done after the training by multiplying each content of the memory locations by a scale factor, n_{max}/n_m, where n_{max} is the number of patterns in the most populous class and n_m the number of patterns in the class being scaled. Since the bit combination (0000) is the most frequently encountered in many chemical patterns but carries the least information, this n-tuple was neither used for training nor classification. This method somewhat reduced the undesirable effect of overtraining.

4.3. Chemical Applications

The original, pure perceptron concept was probably not used for chemical classification problems but the intrinsic idea is found in several other methods. Classification by an adaptive, digital learning

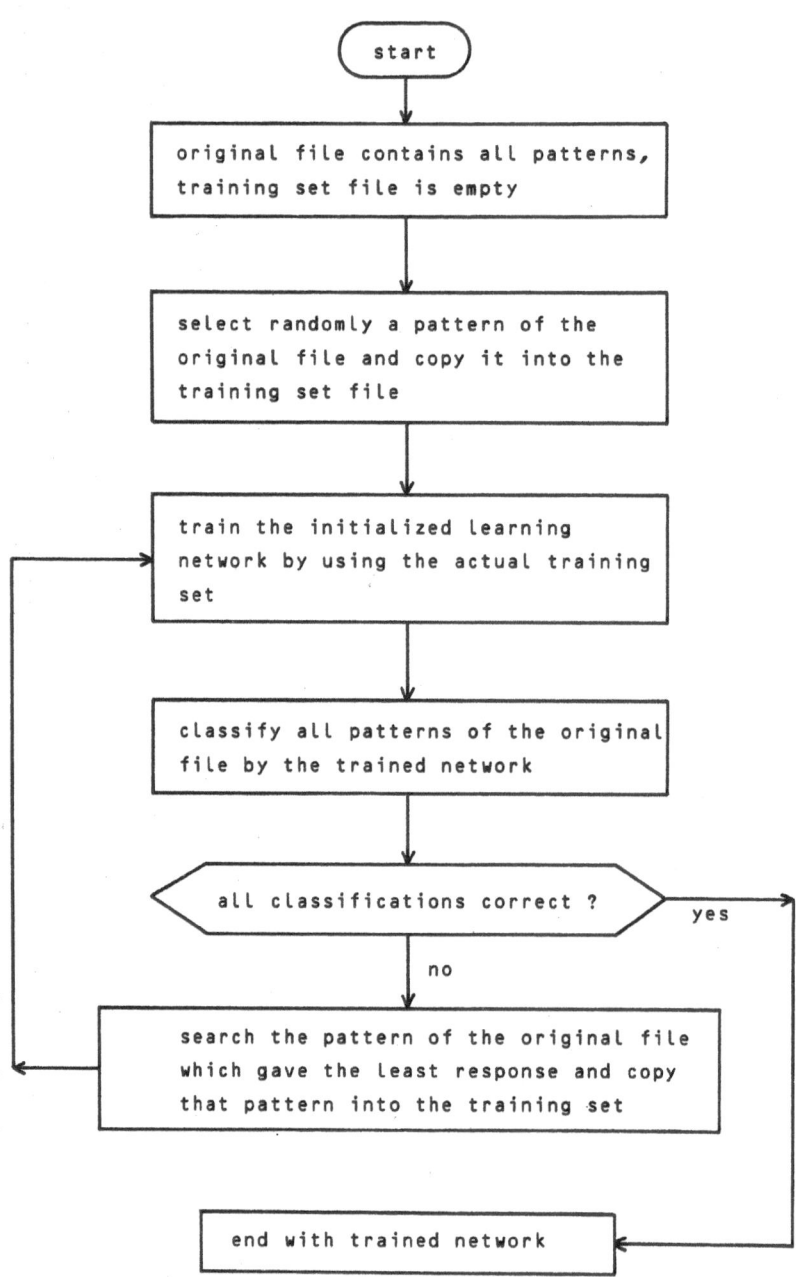

FIGURE 36. Generation of a reduced optimum training set for the adaptive digital learning network [285].

network was applied to the interpretation of mass spectra and some success has been reported even for multicategory problems. Only small differences in the performance of networks have been found when the random combination of features were varied [283 - 285]. Considerably less satisfactory results were achieved for mass spectra by Soltzberg, Wilkins et.al. [280]. The performance of the classifier was very sensitive to the composition of the training set.

The digital learning network is a simple and fast method for the classification of binary encoded patterns. The method suffers especially from the fact that a larger data set gives less satisfactory results. The digital learning network classifier may offer advantages when few carefully selected training patterns are available. Most work has been carried out with random connections of features. Significant improvements may be expected if physically meaningful connections of features are found and realized in a classification network.

5. Parametric Classification Methods

5.1. Principle

Pattern recognition algorithms are often categorized as parametric
or non-parametric. Parametric methods require the knowledge of the
"statistics of the classification problem". If the probability of each
class is known at any location in the d-dimensional pattern space, then
an optimum classification of an unknown pattern can be made by selection
of the "most probable" class at that point. The statistics of the classification problem is estimated by the use of a training set which should
be as large as possible. In practical problems the actual statistics can
never be known exactly because this would require that all possible
measurements had been performed. The available data are never fully representative of a problem and therefore only less than optimum classifications can be achieved.

The probability density functions cannot be stored point by point
because they depend on many (d) variables. Therefore, several parametric
classification methods assume Gaussian distributions and the estimated
parameters of these distributions are used to specify a decision function. Another assumption of parametric classifiers are statistically independent pattern features.

The accuracy of the average risk of a classification can be computed
in principle if the statistics of the data are known. However, the uncertainties mentioned considerably limit the computation of risk in practical applications. Therefore, the usefulness of a parametric method for
a practical classification problem can be examined - as for other classification methods - only empirically.

5.2. Bayes- and Maximum Likelihood Classifiers

The Bayes strategy [376, 389, 396, 403] is a method for making
optimum decisions. For simplicity, the method is described for a binary
classification but it may also be used for multicategory problems.

Let $\bar{x}$ be the pattern vectors. Each pattern vector defines a point with the coordinates x_1, x_2, ... x_d in the d-dimensonal pattern space. The following definitions are necessary for a description of a Bayes classifier.

n : total number of patterns
n_1 : number of patterns in class 1
n_2 : number of patterns in class 2
$p(1) = n/n_1$: a priori probability of class 1
$p(2) = n/n_2$: a priori probability of class 2
$p(\bar{x}|1)$: probability density of class 1 as a function of variables
       x_1, x_2, ... x_d.
$p(\bar{x}|2)$: probability density of class 2 as a function of variables
       x_1, x_2, ... x_d
L(1) : loss associated with misclassification a member of class 1
L(2) : loss associated with misclassification a member of class 2

The classification of an unknown $\bar{x}$ is guided by the following goals:

- Take the class with the largest probability density at location $\bar{x}$.
- Consider the a priori probabilities of the classes to achieve on an
  average maximum success.
- Consider the loss for a misclassification.

The discriminant function for a binary classification is given by equation (78).

$$s(\bar{x}) \quad = \quad \frac{p(1)\ L(1)\ p(\bar{x}|1)}{p(2)\ L(2)\ p(\bar{x}|2)} \quad - \quad 1 \qquad \begin{array}{l} s > 0 : \text{class 1} \\ s \leq 0 : \text{class 2} \end{array} \qquad (78)$$

For a multicategory classification problem with M classes conditional probabilities $p(\bar{x}|m)$ are calculated for each class m.

$$p(\bar{x}|m) \quad = \quad \frac{p(m)\ L(m)\ p(\bar{x}|m)}{\sum\limits_{m=1}^{M} p(m)L(m)p(\bar{x}|m)} \qquad (79)$$

The class with the largest $p(\bar{x}|m)$ is selected.

The Bayes classifier minimizes the total expected loss. The minimum expected risk of a Bayes classification can be calculated if all

functions and variables in equation (79) are known. No classification
method is possible - parametric or non-parametric - with a smaller risk
[152, 374].

The loss associated with misclassification may be equal for all
classes (symmetrical loss function, e.g. L(1) = L(2) = 1) or may be
specified separately for each combination of "computed class" / "actual
class".

Some hesitations may arise with the use of a priori probabilities
of classes during classification of unknowns. If the a priori probabilities are taken from the composition of the training set one maximizes
the total performance of the classifier, assuming for the unknowns the
same a priori probabilities as for the training set. In a practical
classification problem (e.g. "does the molecule contain a carbonylgroup
or not ?") one should usually presuppose equal a priori probabilities for
both classes - otherwise a less frequent class may never be found by the
classifier.

If all classes have equal a priori probabilities, and if the loss
functions are symmetrical, the classification of an unknown pattern $\bar{x}$
needs only the determination of probability densities $p(\bar{x}|m)$ for all
classes m. The maximum value gives the class to which $\bar{x}$ is classified
This method is called a <u>maximum likelihood</u> decision.

5.3. Estimation of Probability Densities

Estimation of the class-dependent probability densities $p(\bar{x}|m)$
is the most important problem in the implementation of a Bayes classifier [403].

A first approach is the estimation of $p(\bar{x}|m)$ for each pattern $\bar{x}$ to
be classified by using known patterns in the neighbourhood of $\bar{x}$. For
this computations the KNN-technique or potential functions may be used
(Chapter 3). The advantage of this approach is that no special assumptions have been made about the form of the probability density function;
the disadvantage is that the whole set of known patterns is necessary
for each classification.

Another approach assumes that the actual probability density can
be approximated by a mathematical function. The patterns of the training
set are used to calculate the parameters of that function.

Often, a d-dimensional Gaussian distribution is used which is defined by
the mean vector of a class and the covariance matrix [380, 389, 391,
396, 403]. Ellipsoidal clusters are well described by this function.
Once the parameters of the Gaussian probability density functions for
all classes are known, the density at any location can be calculated
and an unknown pattern can be classified by the Bayes rule or by the
maximum likelihood method. A binary classification with equal covariance
matrices for both classes can be reduced in this way to a linear classifier [87, 317, 396].

A third approach is the application of the method most frequently
used for chemical applications. The d-dimensional probability density is
computed by a combination of the probability densities of all features.
This method assumes that all features are statistically independent.
The class conditional probability density $p(x_i|m)$ for a feature x_i can
be estimated from a histogram as shown in Figure 37.

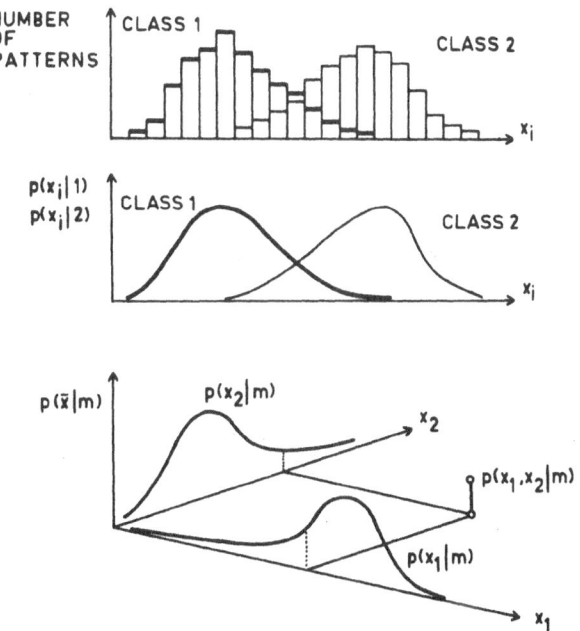

FIGURE 37. Histograms of feature x_i for class 1 and class 2 for the
estimation of the probability densities $p(x_i|1)$ and $p(x_i|2)$. If the com-
ponents of the patterns are statistically independent, the d-dimensional
probability density $p(x_1,x_2|m)$ is given by the product of the 1-dimen-
sional densities of the components.

For an unknown pattern $\bar{x}$ the components x_i yield densities $p(x_i|1)$ for class 1 and densities $p(x_i|2)$ for class 2. The d-dimensional probability density $p(\bar{x}|m)$ for class m is the product of all $p(x_i|m)$.

$$p(\bar{x}|m) \quad = \quad \prod_{i=1}^{d} p(x_i|m) \tag{80}$$

Using the maximum likelihood method (Chapter 5.2) a decision function for a binary classification is defined in equation (81).

$$s \quad = \quad \sum_{i=1}^{d} \lg\frac{p(x_i|1)}{p(x_i|2)} \qquad \begin{array}{l} s > 0 : \text{class 1} \\ s \leq 0 : \text{class 2} \end{array} \tag{81}$$

Instead of two probability density curves for each feature only one function $q(x_i)$ for each feature has to be stored for classifications of unknowns.

$$q(x_i) \quad = \quad \lg\frac{p(x_i|1)}{p(x_i|2)} \tag{82}$$

If $p(x_i|1)$ and $p(x_i|2)$ are approximated by Gaussian distributions, only two parameters have to be stored for each feature. If an approximation by a mathematical function is not possible then all functions $q(x_i)$ must be tabulated for several intervals of x_i (e.g. 3 to 20 intervals). The boundaries of the intervals should be selected such that approximately equal numbers of patterns fall into each interval. Usually, the boundaries are different for each feature and classification of unknowns requires the storage of a large amount of data [248].

This classification method was first applied to chemical problems by Franzen and Hillig [86, 87, 108]. Although many simplifications have been introduced into this maximum likelihood method a considerable computational effort is necessary for the training and application of such parametric classifiers. However, the effort is much smaller if binary encoded patterns are used (Chapter 5.4).

5.4. Bayes- and Maximum Likelihood Classifiers for Binary Encoded

Patterns

In a binary encoded pattern $\bar{x}$ each component x_i has the discrete
value of '0' or '1'. The d-dimensional probability density of a class m
is only defined by d probabilities $p(x_i|m)$ with i = 1, 2, ... d.

$p(x_i|m)$: probability that feature i contains a '1' in patterns of
                class m

$\{1-p(x_i|m)\}$: probability that feature i contains a '0' in patterns of
                class m

Statistical independence of all components is assumed as for other parametric methods. The overall probability (joint probability) that pattern
$\bar{x}$ belongs to class m is given by the product of the probabilities for
all components.

$$p(\bar{x}|m) \quad = \quad \prod_{i=1}^{d} p(x_i|m)^{x_i} \{1-p(x_i|m)\}^{1-x_i} \tag{83}$$

Notice that x_i can be only 0 or 1 in equation (83).

The logarithm of equation (83) gives a maximum likelihood classifier
with a set of decision functions G for all classes m (m = 1, 2, ... M).

$$G(\bar{x}|m) \quad = \quad \sum_{i=1}^{d} x_i \log \frac{p(x_i|m)}{1-p(x_i|m)} \quad + \quad \sum_{i=1}^{d} \log \{1-p(x_i|m)\} \tag{84}$$

Equation (84) is equivalent to the computation of the scalar product of
an augmented pattern vector $\bar{x}$ and a weight vector $\bar{w}_m$.

$$G(\bar{x}|m) \quad = \quad \bar{x} \cdot \bar{w}_m \qquad \bar{w} = (w_{1m}, w_{2m}, \cdots w_{dm}, w_{d+1,m}) \tag{85}$$

$$w_{im} \quad = \quad \log \frac{p(x_i|m)}{1-p(x_i|m)} \qquad i = 1, 2, \ldots d$$

$$w_{d+1,m} \quad = \quad \sum_{i=1}^{d} \log \{1-p(x_i|m)\}$$

The binary encoded pattern $\bar{x}$ $(x_1, x_2, \ldots x_d)$ which is classified by $\bar{w}_m$ must be augmented by an additional component $x_{d+1} = 1$.

Because x_i is 0 or 1 the computation of a scalar product is reduced to a summation of all weight vector components. In this way, a maximum likelihood classification can be rapidly applied.

In a multicategory classification an unknown pattern $\bar{x}$ is assigned to that class m for which $G(\bar{x}|m)$ is a maximum [176, 244, 333, 356, 357].

If the Bayes strategy should be used instead of the maximum likelihood principle the a priori probabilities p(m) of all classes m must be considered. For this purpose, the right side of equation (83) is multilied by p(m). The ratio of Bayes probabilities for a binary classification is given in equation (86) [333].

$$\text{ratio} \quad = \quad \frac{p(1)}{p(2)} \quad \prod_{i=1}^{d} \left[\frac{p(x_i|1)}{p(x_i|2)} \right]^{x_i} \left[\frac{1-p(x_i|1)}{1-p(x_i|2)} \right]^{1-x_i} \tag{86}$$

$$\text{for} \quad p(x_i|m) \quad = \quad 0 \quad \text{or} \quad 1$$

The logarithm of (86) defines a <u>Bayes classifier</u> for binary encoded pattern $\bar{x}$. The discriminant function B is linear and distinguishes between two classes (equation (87)).

$$B(\bar{x}) \quad = \quad \bar{x}.\bar{w} \qquad\qquad \begin{matrix} B > 0 : \text{class 1} \\ B \le 0 : \text{class 2} \end{matrix} \tag{87}$$

$$\bar{w} \quad = \quad (w_1, w_2, \ldots w_d, w_{d+1})$$

$$w_i \quad = \quad \log \frac{p(x_i|1) \, \{1-p(x_i|2)\}}{p(x_i|2) \, \{1-p(x_i|1)\}} \qquad i = 1, 2, \ldots d$$

$$w_{d+1} \quad = \quad \sum_{i=1}^{d} \log \frac{1-p(x_i|1)}{1-p(x_i|2)} \quad + \quad \log \frac{p(1)}{p(2)}$$

$p(x_i|1)$ and $p(x_i|2)$ are the probabilities for the occurrence of a '1' in feature i for class 1 and class 2, resp.. The binary encoded pattern $\bar{x}$ must be augmented by an additional component $x_{d+1} = 1$.

Woodruff, Ritter, Lowry, and Isenhour [355] described an extension of this method using correlation terms.

5.5. A Simple Sequential Classification Method Based on Probability
Densities

Lowry, Marshall and Isenhour [170, 173] introduced a sequential
classification method into chemical applications of pattern recognition.
This algorithm utilizes simplified probability density curves and has
some similarities to parametric methods. The method was called by the
authors "progressive filter network".

Figure 38 shows class-dependent probability density curves for a
pattern feature x_i. All patterns with $x_i < x_{i,min}$ belong to class 1 and
all patterns with $x_i > x_{i,max}$ to class 2. Therefore, patterns which
satisfy one of these conditions can be classified by using only this
single feature. Figure 38 shows possible arrangements of overlapping
probability density curves. In each case, one class can be given immedi-
ately if the value of a feature falls below the minimum or above the
maximum of the overlap. In a training process those features are selected which have the smallest overlap in their probability density curves
and therefore classify a maximum number of patterns correctly.

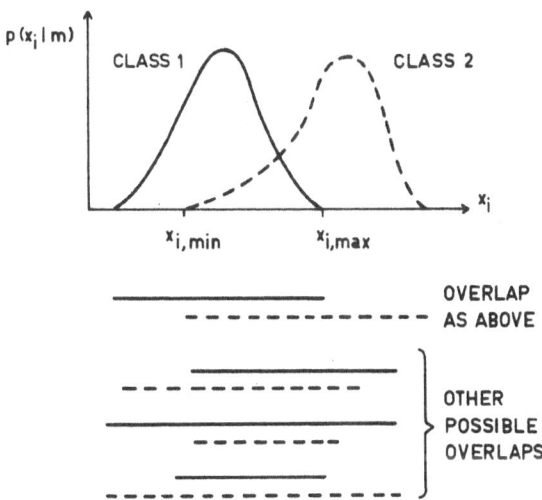

FIGURE 38. Probability density curves for feature i in a two-class
problem. Overlap of the curves is possible in four different arrangements.

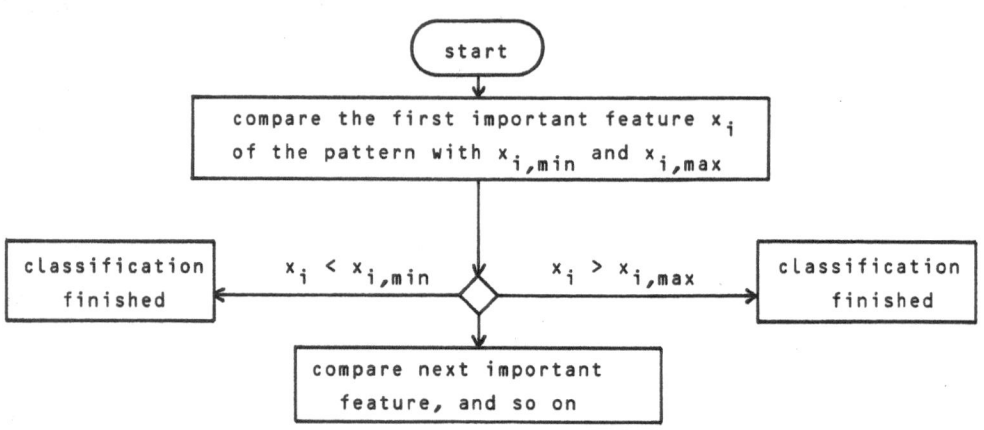

FIGURE 39. Progressive filter network for a binary classification of an unknown pattern $\bar{x}$ [173].

All selected features are put into a decision network in descending order of classifying ability (or, if known, in descending order of physical importance) as demonstrated in Figure 39. The boundaries $x_{i,min}$ and $x_{i,max}$ of each feature and the class assignment outside the boundaries are included in the decision process. If a considered feature of an unknown pattern lies outside the boundaries, the class is assigned and classification is terminated. Otherwise, the pattern "falls" into the next comparison.

The algorithm is simple and fast. Difficulties arise in the determination of the boundaries if large data sets are applied. The feet of the probability density curves spread out and minimum and maximum values of the overlapping region become useless.

A much more sophisticated classification method with some similarities to the progressive filter network has been proposed by Meisel et. al. [205]. A set of decisions for a given classification problem is combined in a decision tree. Each terminal node of the tree indicates the probabilities for class 1 and 2.

5.6. Chemical Applications

A statistical meaningful estimation of probability densities requires very large data sets. Therefore, chemical applications of parametric classification methods always include assumptions which are often
not fulfilled or cannot be proved. A severe assumption is the statistical
independence of the pattern components which is certainly often not satisfied. Generation of new independent features is usually too laborious
(Chapter 10).

Successful and economical applications of maximum likelihood
classifiers have been reported for binary encoded infrared spectra [356]
and nuclear magnetic resonance spectra [357]. An extensive examination of
a maximum likelihood algorithm for the interpretation of mass spectra was
made by Franzen et.al. [86, 87, 108] and others [200, 248]. Approximation
of the probability density by a mathematical function has found up to
now only little interest [317].

6. Modelling of Clusters

6.1. Principle

A usual pattern recognition problem in chemistry is to determine whether or not a compound is of a given type. The assumption often made that all patterns belonging to the same class form a distinct cluster is often unrealistic. An <u>asymmetric situation</u> occurs if one class does not form a proper homogeneous group [7]. In Figure 40 the patterns of class 1 form a rather compact cluster while class 2 is scattered throughout the whole pattern space. In a practical example, class 1 may correspond to "good" samples and class 2 to "bad" samples (equivalent to outliers whose features differ markedly from standard values). In such a case, it is useful to construct a model of class 1. The model is utilized either for the classification of unknown patterns or for the interpretation of the data structure. A confidence region can be constructed around the class model. Patterns outside this region are considered to belong to the "unstructured" class.

The appropriate tool for the construction of models is factor analysis and principle component analysis [189]. An introduction to these statistical methods is beyond the scope of this book and therefore only a brief discussion about modelling of clusters is given here.

6.2. Modelling by a Hypersphere

A very simple model of class 1 in Figure 40 is a hypersphere around the centre of gravity. Classification of an unknown pattern is made by comparing the distance D to the centre of gravity with a critical radius r_c. A decision criterion y can be defined as

$$y = r_c - D \qquad \begin{array}{l} y > 0 : \text{class 1} \\ y < 0 : \text{class 2} \end{array} \qquad (88)$$

The determination of r_c is facilitated if class-dependent probability

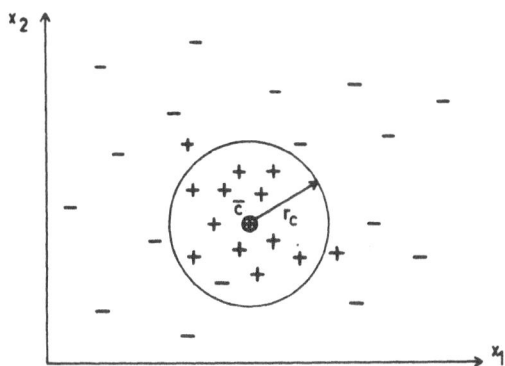

FIGURE 40. Modelling class 1 (+) by a circle (d-dimensional hyper-sphere) around the centre of gravity $\bar{c}$. Within a critical radius r_c almost only patterns of class 1 are found.

densities are calculated as a function of the distance to the centre of gravity. The determination of the critical radius depends on the require-ments of classification. If a high recall for class 1 is required (t.m. no members of class 1 should be missed even if the precision of the classification may be lowered) a relatively large critical radius is necessary.

6.3. SIMCA - Method

A pattern recognition method based on the modelling of each separate class by a principal component model was developed by Wold [342]. This excellent and efficient "soft independent modelling of class analogy" or "statistical isolinear multicategory analysis" was successfully applied to several chemical problems. A detailed description of SIMCA requires more knowledge of mathematics and statistics [428, 429, 431] than many other pattern recognition methods. Therefore, only the principle of the method will be given here; detailed descriptions are found in [7, 189, 342, 344 - 346, 349].

In the SIMCA method an a-dimensional hyperplane is fitted to each class (Figure 41). A confidence plate is constructed around this plane. The width and length of the plate is limited to the basis of the distribution of patterns along the corresponding dimensions. This gives a separate <u>hyperbox</u> ("envelop") for each class. The parameters of a hyperbox are calculated from a training set of patterns by principal component analysis. The axes of a hyperbox are mutually orthogonal and correspond to directions of maximum variance. The main difference between SIMCA and the factor analysis is that the latter usually assumes a single model to be valid over all data. A hyperbox is a local and soft model because it is limited to a single class and of approximate nature.

This concept allows a mathematical description of a single class of patterns. Outliers with deviating features lie outside the box. SIMCA gives for each unknown pattern a probability for each class. Patterns with very low probabilities for all classes probably indicate "a new kind".

An interclass distance is measured as the distance between two hyperboxes relative to their thickness. This measure is used to characterize a classification problem.

SIMCA is not only applied to classification of unknowns but also to the prediction of one or several "external properties" of unknown patterns.

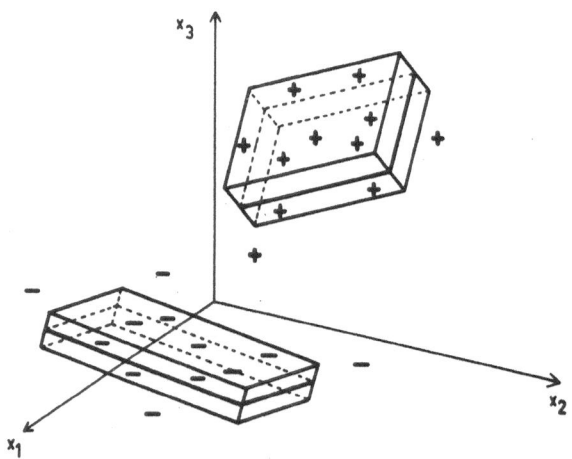

FIGURE 41. In the SIMCA method a hyperbox is constructed for each class.

Chemical applications of SIMCA have been summarized by Wold et.al. [7, 349]; they include :

- classification of infrared and ultraviolet spectra of carbonyl compounds [344],
- classification of ^{13}C-nuclear magnetic resonance spectra [266],
- investigations of relationships between chemical structure and biological activity [81, 82, 219],
- recognition of oil spills [80],
- classification of blood samples from welders [301].

Application of SIMCA to problems of linear free energy relationships are summarized in [350].

7. Clustering Methods

7.1. Principle

If an assignment of classes to patterns is not evident, then
unsupervised learning methods are often helpful. Methods of finding
clusters in a multidimensional pattern space are used to find natural
classes in a data set. Such methods are not trivial and always contain
heuristic and arbitrary elements. Subjective parameters are necessary
to control the size, shape and number of clusters for a certain problem. Different representations of the data often give different clusters.

The purpose of cluster analysis is to find chemical meaningful
classes which correspond to clusters in the pattern space (and are therefore recognizable by classifiers). A cluster analysis may show the complexity of a classification problem.

The guiding principle of cluster analysis is that the more similar
two patterns are, the closer they lie together in the pattern space.
(Several distance measurements are defined in Chapter 2.1.9.) There are
two general approaches to cluster analysis. In an agglomerative method
each pattern is initially considered to be a single cluster. Larger
clusters are built by merging smaller clusters. The other philosophy is
a divisive method. This process begins with all patterns in one large
cluster. Smaller clusters are separated by successive splitting of the
existing structure. Numerous mathematical clustering methods have been
described [189, 272, 375, 380, 408].

Generally, useful techniques for finding clusters are "hierarchical
clustering" (Chapter 7.2) and "minimal spanning tree" (Chapter 7.3).

Another method of cluster analysis involves projection of the
d-dimensional pattern space onto 2 or 3 dimensions. The clusters may
then be distinguished visually. Methods of projection are described in
Chapter 8.

7.2. Hierarchical Clustering

Hierarchical clustering is a so-called "SAHN-technique" (an abbreviation for sequential, agglomerative, hierarchical, non-overlapping);

- sequential because a sequential grouping algorithm is used,
- agglomerative because the procedure starts from individual patterns which are merged to clusters,
- hierarchical because there are always less clusters at each ascending level of clustering,
- non-overlapping because patterns cannot be simultaneously members of several classes.

First the distance matrix is calculated (distances between pairs of pattern points). The smallest distance is searched and the two corresponding patterns are combined to a new point half-way between (centre of gravity). The number of patterns is thereby reduced by one (Figure 42). The distance matrix is recalculated for the reduced set and again the closest pair of points is seeked. This process is repeated until all patterns have been combined. The algorithm can treat all clusters equally or weight centres of gravity calculations according to the number of patterns in each cluster.

Instead of a distance measure a similarity measure was proposed by Kowalski and Bender [153]. The similarity S_{jk} between patterns $\bar{x}_j$ and $\bar{x}_k$ is defined by equation (89).

$$S_{jk} = 1 - D_{jk} / max(D_{jk}) \tag{89}$$

where D_{jk} is the Euclidean distance between $\bar{x}_j$ and $\bar{x}_k$ and $max(D_{jk})$ represents the largest distance in the data set. Values for S_{jk} range from zero (least similar patterns) to unity (identical patterns).

The result of the clustering is usually presented in a dendogram (Figur 43). The dendogram shows the connections of pattern points and on the ordinate the actual minimum distance for each clustering step.

Examination of a dendogram in order to separate clusters can be made in different ways either by an algorithm or by the scientist.
1. The desired number of clusters can be specified and the dendogram shows the composition of the clusters.
2. A predetermined maximum distance (or minimum similarity value) is set and clustering is stopped when this distance is reached.

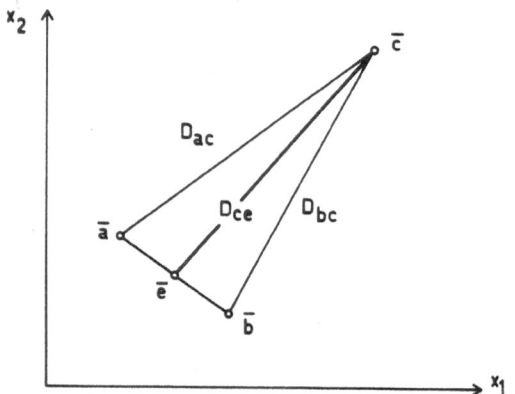

FIGURE 42. Hierarchical clustering. Points $\bar{a}$ and $\bar{b}$ are closest in the original data set ($\bar{a}$, $\bar{b}$, $\bar{c}$) and therefore merged to a new point $\bar{e}$. The new distance D_{ce} is obtained by averaging the distances D_{ac} and D_{bc}.

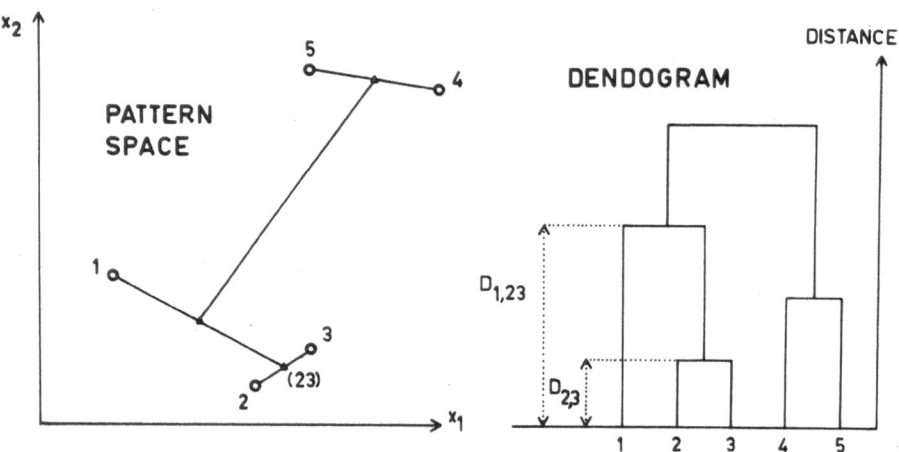

FIGURE 43. Hierarchical clustering. Two-dimensional example with five pattern points.

3. The rate of change in the distances (or similarity values) needed to produce the next clustering can be checked. The procedure stops when a certain value (e.g. 5 %) is exceeded.

   Some arbitrariness is unavoidable in the separation of the clusters. In the example shown in Figure 43 one may recognize two clusters (1, 2, 3) and (4, 5) or four clusters (1), (2, 3), (4) and (5). Actual problems from chemistry are certainly still more complicated.

   Hierarchical clustering shows especially the local structure of a data set.

7.3. Minimal Spanning Tree Clustering

   At the start of this method all pattern points are connected together. The mathematical requirement for this connection is a minimum total length of all line segments (Figure 44). Such a connection network is called a minimal spanning tree; it may contain branches but must not contain circuits.

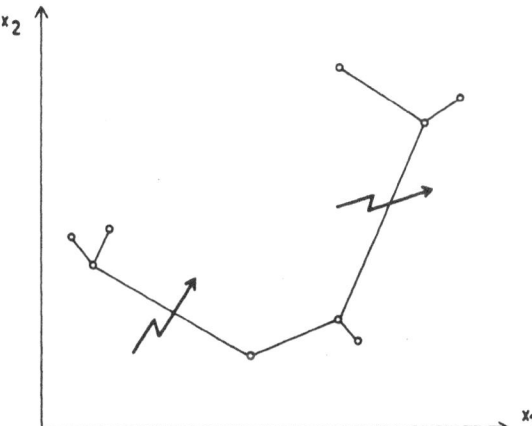

FIGURE 44. Clustering by a minimal spanning tree. The tree is broken into clusters by cutting all segments that are longer than a predetermined value.

The problem to find the shortest path through a set of points can be formulated as the "travelling salesman problem". Suppose a salesman has to visit customers in several towns and he looks for the shortest way through all towns. An exact description of the problem and the solution requires some knowledge in graph theory [189, 240, 386, 408]. The calculation is very laborious for more than 100 patterns.

For cluster analysis also simpler algorithms were proposed which find a short but not necessarily the shortest spanning tree [268].

Various methods are possible to break the minimal spanning tree into clusters [240, 408]. In one of the methods all line segments are cutted that are longer than a length supplied by the researcher. This method is particularly effective in the detection of outlying patterns.

7.4. Chemical Applications

Probably the most impressive historical example of cluster analysis in chemistry is the discovery of the Periodic Table of elements – although no "modern mathematical clustering methods" have been used.

Extensive applications of clustering techniques have been reported for the grouping of liquid phases for chromatography [83, 145, 189, 240] and for the grouping of elements analyzed in environmental samples [90, 111]. Clustering methods have been used to generate an ordered list of sulfur-containing compounds according to their resemblence of mass spectra [106].

Cluster analysis techniques are usually limited to rather small data sets with 20 to 100 patterns.

8. Display Methods

8.1. Principle

The goal of display methods is to visualize the data structure of
a d-dimensional pattern space by a 2- or 3-dimensional representation.
The human is the best pattern recognizer in a 2- or 3-dimensional world,
thus the scientist may recognize clusters and assign classes to unknown
patterns according to their nearness to known patterns.

Display methods try to conserve all distances between patterns in
the hyperspace. Clearly, this cannot be done exactly during a reduction
of the dimensionality. There are two basic approaches to the display of
a d-dimensional space (d>2) in a 2-dimensional plane : linear or non-
-linear. A pattern point $\bar{x}$ is defined by d coordinates x_1, x_2, ... x_d.
In a 2-dimensional representation the same pattern point has 2 coordinates y_1 and y_2 which can either be linear or nonlinear combinations
of the original coordinates x_i. Linear methods are often called
projections and nonlinear methods are called mappings.

The preservation of the local distances can be evaluated by using
the K-nearest neighbour classification (Chapter 3) and comparing the
results for the d-dimensional patterns and for the 2-dimensional representations.

Numerous display methods have been proposed in the literature [377,
380, 395]; most of them are very cumbersome. An excellent review from
the viewpoint of chemical applications was written by Kowalski and Bender
[155].

8.2. Linear Methods

The simplest way of a linear projection is to select a pair of
variables (features) and to generate a 2-dimensional plot. For a set of
d features there are d(d-1)/2 independent "variable by variable plots".
These plots are seldom useful when d is large. Selection of the two
"most important" features (see Chapter 10) may give valuable information
on simple classification problems.

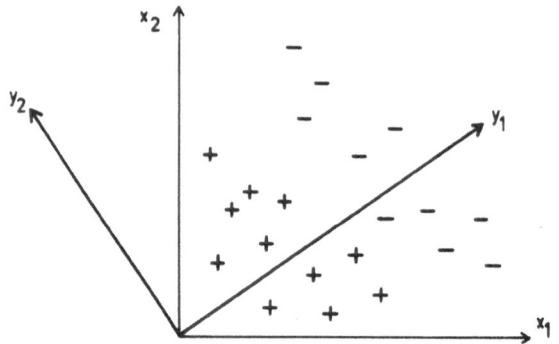

FIGURE 45. Projection of the 2-dimensional pattern space onto axis x_1 or x_2 destroys the structure of the clusters. An appropriate rotation of the axes defines new coordinates y_1 and y_2 which are linear combinations of the original coordinates. Projection onto the y_1-axis preserves the structure of the clusters.

Figure 45 shows that a rotation of the coordinate axes may improve variable by variable plots significantly. This method is very useful if an, interactive computer graphics terminal is available and if the dimensionality of the pattern space is not too large. A multidimensional rotation can be considered as a series of $d(d-1)/2$ two-dimensional rotations.

A very useful and in some sense optimum multidimensional rotation is given by a <u>Karhunen-Loeve</u> transformation. The Karhunen-Loeve (KL) method creates new variables y_k as linear combinations of the original variables x_i. The new variables are ordered uniquely. The first new variable y_1 contains the greatest amount of variance (t.m. y_1 corresponds to the direction of greatest spread through the pattern space). Each successive new variable contains the next greatest amount of the residual variance in a direction orthogonal to the previous axes. The transformation is optimal in the sense that variance is preserved. The first two (or three) new variables are used to generate a plot of the d-dimensional pattern space. This plot contains more information than any other combination of two original variables.

The mathematics of the KL method cannot be described here in detail;
see [428, 429, 431]. The method starts by calculating the means m_i
for all d variables x_{ij}, where x_{ij} is the i-th variable of pattern $\bar{x}_j$.
Next the covariance matrix is generated. Each element c_{ih} of this matrix
compares two variables i and h and is given by equation (90); n is the
number of patterns.

$$c_{ih} = \sum_{j=1}^{n} (x_{ij} - m_i)(x_{hj} - m_h) \tag{90}$$

Next the eigenvalues and eigenvectors of the covariance matrix are calculated. The two eigenvectors corresponding to the two largest eigenvalues represent the two new axes with greatest variance and are used for
a display (eigenvector plot). For each new axis the percent variance retained by the projection can be given. This value is useful for determining the reliability of the interpretation of a certain display (e.g.
a projection may contain 80 % of the total variance of the data).

The KL method as described above maximizes the sum of all squared
distances between pattern points in the projection. No information about
the membership to certain classes is considered - it is an unsupervised
method. A supervised approach is a class-dependent projection [395].
In this version the search for an optimum projection plane is guided by
the goal of maximum separation of the known classes. One approach is to
maximize the sum of all squared distances between patterns of different
classes. Another approach is to minimize the sum of squared distances
within each class. Also a combination of both methods has been proposed
[248].

Linear display methods have the advantage that a recalculation of
the transformation matrix is not necessary for a projection of new
patterns. Unknown patterns can be easily projected onto the display
of a training set and classified according to their nearness to known
patterns.

8.3. Nonlinear Methods

These methods produce displays with coordinates y_k that are not linear combinations of the original coordinates x_i in the d-dimensional pattern space [380].

A simple approach utilizes the distances between two particular points in the d-dimensional space as coordinates in a 2-dimensional display. In a binary classification problem one may use the centres of gravity of the classes as reference points. The method preserves some geometric structur and puts heavy weight on class separability [155, 381]. A similar method for the display of multivariate chemical data was proposed by Lin and Chen [169] and Drack [73, 74]: three pairs of reference points in the d-dimensional space are employed and new coordinates are calculated so as to give the same ratio of distances to a pair of reference points.

The most popular nonlinear display method was proposed by Sammon [401] and is called "nonlinear mapping" (NLM). The technique seeks to conserve interpoint distances. Let be

D_{ju}^* : distance in the d-dimensional space between pattern $\bar{x}_j$ and $\bar{x}_u$; these distances are calculated from the given coordinates of the patterns and therefore considered as constants;

D_{ju} : distance in a 2-dimensional display between pattern $\bar{x}_j$ and $\bar{x}_u$;

y_{1j}, y_{2j} : new coordinates for pattern $\bar{x}_j$ in the 2-dimensional display;

n : number of pattern points.

An ideal display would have $D_{ju} = D_{ju}^*$ for all pairs of patterns. Since this cannot be done exactly an error e is defined (equation (91)).

$$e = \sum_{\substack{j=1 \\ u>j}}^{n} (D_{ju}^* - D_{ju})^2 (D_{ju}^*)^{-k} \tag{91}$$

The parameter k is used to emphasize different aspects of the data structure. Positive k values emphasize the global structure and negative values (e.g. k = -1) the local structure; k = 2 corresponds to an equal weighting of small and large distances.

Minimizing the error e is carried out by adjusting all 2n variables y_{1j} and y_{2j}. This optimization is complicated for a large set of patterns and expensive in computer time. Modifications of the NLM method were proposed by Kowalski and Bender [155] and by Rossell and Fasching [247]. These versions start the optimization with an eigenvector plot. Optimization is conducted by a gradient method or by the simplex method.

Nonlinear mapping has an analogy in physics. Suppose each of the pattern points in the d-dimensional space is connected to every other point by springs. In the original configuration all of the springs are completely at rest; the total tension is zero. NLM compresses the whole network of points and springs into fewer dimensions with the goal of minimum total tension. The error function (91) can be used as an estimate of the total energy in the springs [155, 199].

Several other graphical techniques have been developed for a visual representation of multivariate data in two dimensions [377]. Chernoff [373] described a curious method in which each pattern vector is represented by a cartoon face; the features of the face are governed by the vector components. It is assumed that the human can easily recognize similar faces corresponding to a cluster of pattern points.

8.4. Chemical Applications

Display methods form a very useful branch of pattern recognition. They allow the scientist to have an approximate view of the d-dimensional space. The scientist can examine a data structure and then select another appropriate pattern recognition method. A fruitful "cooperation" may be expected between the computer (an automatized procedure presents the data in a new and clear display) and the scientist (whose pattern recognizing capabilities cannot be obtained by an algorithm).

Disadvantages of these methods are the necessary large mathematical and computational efforts. One also must never forget that a 2-dimensional display of a d-dimensional space is sometimes only a very crude approximation of the real configuration.

Linear projections and nonlinear mappings have been used for several chemical classification problems [153], e.g. classification of archaeological data [162] or interpretation of analytical data in environmental research [90]. Most of the display methods have also been demonstrated by synthetical data.

9. Preprocessing

9.1. Principle

The typical steps during an application of pattern recognition methods for a chemical classification problem are:

- Empirical selection of measurements that are expected to be related to the classification problem.
- Preprocessing of the raw data and selection of relevant features.
- Training of a classifier.
- Evaluation of the trained classifier.
- If the performance of the classifier is not good enough then modifications on the first three steps are necessary.

Preprocessing is numerically operating on the raw data (measurements) in order to increase the usefulness for pattern recognition methods. Preprocessing changes the structure of the arrangement of pattern points in the pattern space. If the measurements have different magnitudes a scaling is necessary (Chapter 9.2). Weighting methods determine which of the features are useful for a separation of given classes (Chapter 9.3). Elimination of meaningless features is described in Chapter 10.

At the moment no general theory exists about an optimum preprocessing for a certain classification problem. Therefore, a very great number of methods have been proposed [387]. The usefulness of a preprocessing method can usually only be judged empirically by an objective evaluation of the corresponding classifier. Justification of preprocessing methods for unsupervised learning is almost impossible.

9.2. Scaling

Measurements with a great absolute magnitude have a more pronounced effect on the classification result than other measurements. In a scaling process the measurements are adjusted to have some common property. Scaling is especially important for multisource data (Chapter 1.2) which are combined in a single vector.

A simple and well known scaling method is <u>range scaling</u>.

$x_{i,old}$: original component i of a pattern vector
$x_{i,new}$: adjusted component i
$x_{i,max}$: largest value for $x_{i,old}$ in the data set
$x_{i,min}$: smallest value for $x_{i,old}$ in the data set

$$x_{i,new} = \frac{x_{i,old} - x_{i,min}}{x_{i,max} - x_{i,min}} \qquad (92)$$

Range scaling has the disadvantage that if there is one extreme value all the other values become approximately the same.

A very useful scaling method is <u>autoscaling</u>. The measurements are adjusted so that they each have a mean of zero and unit variance.

$$x_{i,new} = \frac{x_{i,old} - m_{i,old}}{v_{i,old}} \qquad (93)$$

$m_{i,old}$: mean of all original components $x_{i,old}$
$v_{i,old}$: variance of original components $x_{i,old}$

After autoscaling all pattern components have an equal weight and therefore an equal effect on classification.

9.3. Weighting

Weighting methods determine the importance of the scaled features for a certain classification problem. Weighting can only be used for supervised learning. Weighting factors are calculated or estimated by
- the statistics of the data,
- evaluation of the classifier,
- chemical knowledge.

Automatic weighting methods are used for feature selection and described
in Chapter 10. Weighting methods attempt to increase the separation between the defined classes. Some precaution is necessary because any
feature that even has slight (perhaps random) separating ability is
emphasized.

9.4. Transformation

Instead of the raw or scaled measurements a mathematical function
of them may be used. Several classification problems in chemistry have
been facilitated by a reduction of the dynamic range. This can be effected by the elementary functions $\sqrt{x}$, log x, log(x+const), etc.

A severe simplification of the data is realized by binary encoding :

$$x_{new} = 1 \quad\quad \text{if } x_{old} \geq \text{threshold} \quad\quad\quad (94)$$
$$x_{new} = 0 \quad\quad \text{if } x_{old} < \text{threshold}$$

9.5. Combination of Features

New features can be generated by a linear or nonlinear combination
of some or all original measurements. Correlations between features can
be considered by the calculation of cross terms which are added to the
pattern vector or may substitute less important features. For a pattern
with d components d(d+1)/2 binary cross terms (e.g. $\sqrt{x_i x_h}$) exist.
Selection of the most important cross terms can be performed by the same
methods as used for feature selection [124, 160, 391]. An autocorrelation function was proposed for the classification of nuclear magnetic resonance spectra [161].

A danger associated with the addition of nonlinear features is the
fact that any set of patterns (even randomly generated) can be separated
into randomly selected classes by adding features [148].

Generation of new orthogonal features can be carried out by a principal component analysis. A Karhunen-Loeve transformation (Chapter 8.2)
gives new non-correlated vector components.

Another method for the calculation of orthogonal features for a given set of patterns is called "SELECT". This method was successfully applied by Kowalski and Bender [158] to chemical classification problems. In this method, the most important original feature is taken as the first new feature. Then, the correlations of the first feature with all other features are eliminated and the most important remaining feature is taken as the next new feature. In this way, subsequent new independent features are generated. Each selected feature retains its identity minus the correlations of the previously selected features. Therefore, the later the feature is selected, the less of its original identity is retained.

For spectral patterns Fourier-, Hadamard-, and Walsh- transformations have been applied to generate new features [118, 154, 335]. Some success of these transformations has been reported for nuclear magnetic resonance spectra [29, 337], and for mass spectra [122, 134, 135, 148, 154, 319]. For non-spectral data such transformations do not seem meaningful and have not been shown to be advantageous.

An heuristic combination of original features may also be used to generate new features; however, such methods are extremely problem-dependent. Bender, Kowalski and Shepherd [16, 17] investigated several possibilities for the representation of mass spectra; means, moments, number of peaks, histograms, etc. were successfully used as new features.

10. Feature Selection

10.1. Principle

The objects (or events) of a data base are characterized by a set
of measurements (features). For pattern recognition purposes the number
of features should be as small as possible for two reasons.

First, features which are not relevant to the classification problem
should be eliminated because they may disturb the classification or at
least enlarge the computational work. For the same reason, correlations
between features should be eliminated.

Second, the number (d) of features must be much greater than that (n)
of patterns. Otherwise, a classifier may be found that even separates
randomly selected classes of the training set. A ratio of $n/d > 3$ is
acceptable, $n/d > 10$ is desirable (Chapter 1.6, details in Chapter
10.4). This second reason forces the user of pattern recognition methods into feature selection. In almost all chemical applications of
pattern recognition is the number of original raw features too large
and a reduction of the dimensionality is necessary.

The problem of feature selection is to find the optimum combination of features. Most methods only look for the best individual
features. An exhaustive search [175, 388] for the best subset of features
is only possible for very small data sets.

Another strategy is the generation of a group of new features (e.g.
linear combinations of the original features) as described in Chapter
9.5. Features which are essential for a classification are often called
"intrinsic features".

Some feature selection methods may be used not only for supervised but also for unsupervised learning; however, most methods require
the knowledge of the class membership for the training set patterns.

One group of feature selection methods uses the statistics of the
data (means, variances) to select the most important features. The
features are ranked according to their importance and less important
features are discarded.

Another group of feature selection methods try to determine the
influence of each feature on a certain classification method. Features

which contribute only little to the computation of a classifier are eliminated. This method - although frequently used - is dangerous if the ratio n/d for the original data set is not large enough. One must not start with an over-determined data set and use the results of classification to lower the dimensionality [290].

Although feature selection has been one of the most intensively investigated areas of pattern recognition, no general theory exists up to now. Therefore, the chemist should never forget the chemical background of his classification problem. A feature selection based on chemical knowledge may often be much more effective than mathematical methods.

10.2. Feature Selection by Using Data Statistics

Each original pattern vector $\bar{x}_j$ contains d components (features) x_{ij} (i = 1, 2, ... d); n is the number of patterns in the data set (j = 1, 2, ... n).

10.2.1. Variance Weighting

One may assume that features with a great variance are more important for a classification than features with a small variance. For a class-independent weighting, variance v_i of feature i is given by equation (95).

$$v_i = \frac{1}{n-1} \sum_{j=1}^{n} (x_{ij} - m_i)^2 \qquad (95)$$

m_i is the mean of feature i.

$$m_i = \frac{1}{n} \sum_{j=1}^{n} x_{ij} \qquad (96)$$

Features with the smallest variances are discarded.

Weighting for a supervised learning uses the ratio of an "inter-class variability" and an "intraclass variability" [148, 153]. The method is applied after autoscaling (Chapter 9.2).

The interclass variability $V_{inter,i}$ of feature i is calculated over all two-pattern combinations $\bar{x}_j$ and $\bar{x}_k$ not in the same class.

$$V_{inter,i} = \sum_{j,k} (x_{ij} - x_{ik})^2 \qquad (97)$$

The intraclass variability $V_{intra,i}$ of feature i is calculated by the same equation (97) but for all two-pattern combinations $\bar{x}_j$ and $\bar{x}_k$ with both patterns in the same class. A weight g_i is calculated for each of the features.

$$g_i = \frac{V_{inter,i}}{V_{intra,i}} \qquad (98)$$

Lager values of g_i indicate the relative importance of feature i.

10.2.2. Fisher Weighting

The Fisher ratio F_i of feature i for a binary classification prob-lem is given by equation (99) [376, 416].

$$F_i = \frac{(m_{i1} - m_{i2})^2}{v_{i1} + v_{i2}} \qquad (99)$$

The mean values m_{i1}, m_{i2} and the variances v_{i1}, v_{i2} have to be calcu-lated separately for patterns of class 1 and class 2. The largest values of F_i correspond to the "most important" features which best reflect the differences between the two classes.

Fisher ratios have a great dynamic range variing from zero if the classes have the same mean to some large value for well separated classes.

Instead of the Fisher ratio an easier to calculate weight was pro-posed by Coomans et.al. [55] (equation (100)).

$$g_{i,\text{Coomans}} = \frac{|m_{i1} - m_{i2}|}{\sqrt{v_{i1}} + \sqrt{v_{i2}}} \tag{100}$$

10.2.3. Use of Probability Density Curves

Complete information on the importance of a single feature i for the separation of classes is shown by class-conditional probability density curves (Figure 46). A small overlap of both curves indicates a high discrimination power of that feature. The overlap may be measured by the common area, by the Bhattacharyya coefficient, by the transinformation, or some other criteria (Chapter 11.6 and 11.7).

Franzen and Hillig [87] used the absolute difference of the class-conditional means $|m_{i1} - m_{i2}|$ as a criterion for feature selection. Similar methods were used by Mathews [192] and by Pichler and Perone [228].

Difficulties arise if a high probability exists for zero values; therefore, a feature selection method was proposed [42, 165] that utilizes the features with most frequently encountered non-zero values (Chapter 10.2.4).

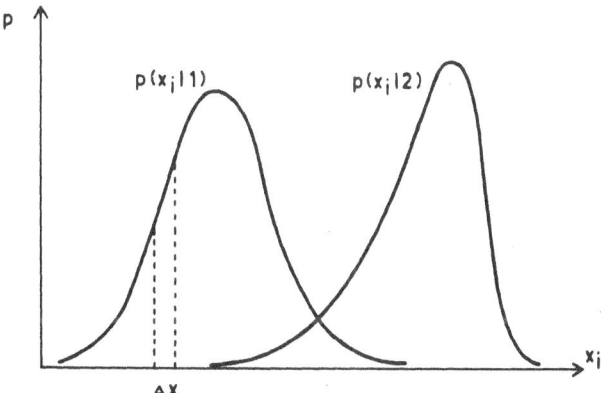

FIGURE 46. Probability density curves of feature i for a binary classification problem. $p(x_i|m)\Delta x$ is the probability that a pattern belonging to class m has the value of feature i in the interval Δx.

10.2.4. Methods for Binary Encoded Patterns

If all features are binary encoded (x_i = 0 or 1) some simplifications and specialities exist. One possible feature selection method
determines those features which have maximum variance among the a posteriori probabilities as calculated by the Bayes rule [170, 171, 353].

The a posteriori probability $p(m|x_i=1)$ of a particular pattern be-
longing to class m under the condition that feature i has the value 1 is
given by equation (101).

$$p(m|x_i=1) \quad = \quad \frac{p(x_i=1|m)\ p(m)}{p(x_i=1)} \tag{101}$$

$p(x_i=1|m)$: probability of a '1' in feature i given that the pattern
belongs to class m

$p(m)$: probability of a pattern belonging to class m

$p(x_i=1)$: probability of a '1' in feature i for all patterns

In a multicategory classification problem with m_{max} classes, a set
of m_{max} a posteriori probabilities exists for each feature. If the probabilities for a particular feature are similar, then this feature does
not contribute to the classification. Those features with the smallest
variances among the m_{max} a posteriori probabilities are eliminated from
the data. The probabilities are approximated by a training set.

Lowry and Isenhour [170, 171] expanded this method by introducing
a cost factor for each feature. The cost factor of feature i is proportional to the number of times a '1' appears in feature i. The weight of
feature i is calculated as the product of the corresponding cost factor
and the variance (among the m_{max} a posteriori probabilities). The cost
factor lowers the importance of those features with only a few '1's and
randomly high variance. This method was successfully applied to a selection of the most important wavelengths in binary encoded infrared spectra.

Another approach [244] to feature selection uses information theory. (A short introduction to the information theory is given in Chapter
11.6.1). The average amount of information on a feature to distinguish
patterns of class 1 from patterns of class 2 is the mutual information
$I_{mutual,i}$.

$$I_{mutual,i} \quad = \quad H(x_i) \ - \ H(x_i|m) \tag{102}$$

$H(x_i)$ is the entropy of feature i over all patterns and is given by equation (103).

$$H(x_i) \quad = \quad - p(x_i=1) \; ld\{p(x_i=1)\} \quad - \quad p(x_i=0) \; ld\{p(x_i=0)\} \qquad (103)$$

$H(x_i|m)$ is the average conditional entropy and is defined for a binary classification problem by equation (104).

$$H(x_i|m) \quad = \quad p(1) \; H(x_i|1) \quad + \quad p(2) \; H(x_i|2) \qquad (104)$$

p(1) and p(2) are the probabilities of class 1 and 2 in the training set. The conditional entropies $H(x_i|m)$ for equation (104) are calculated by inserting the conditional probabilities $p(x_i|1)$ and $p(x_i|2)$ into equation (103) for each class separately. $H(x_i|m)$ is the average information for the distinction of patterns of class m.

Those features with the largest values of the mutual information are the ones that should be most useful for classification. The square root of the mutual information was found to correlate quite well with the recognition ability of maximum likelihood classifiers in an application to infrared spectra [244].

Difficulties arising in applications of feature selection methods to binary encoded patterns have been discussed by Miyashita et.al. [213].

10.2.5. Elimination of Correlations

Most of the feature selection methods neglect second-order effects. If two features are highly correlated both features will be selected if they are valuable for the classification - although the second feature contains the same information as the first one. Optimal methods for the elimination of correlations generate new orthogonal features; this can either be effected by the Karhunen-Loeve method (Chapter 8.2) or by the SELECT method (Chapter 9.5 [158]).

A simpler - but not necessary less laborious - method is the calculation of the correlation coefficients for each possible pair of features in the patterns belonging to a certain class. Pairs of most correlated features are combined or one of them is deleted [55, 199].

A feature selection method - called attribute inclusion algorithm [365], ("attribute" is equivalent to feature) - for binary encoded patterns was applied to chemical problems by Schechter and Jurs [118, 260].

10.3. Feature Selection by Using Classification Results

10.3.1. Evaluation of the Components of a Single Weight Vector

Once a weight vector $\bar{w}$ is trained for a given classification problem one may delete those components with the smallest absolute values $|w_i|$. After elimination of the corresponding features in all patterns of the training set a new weight vector is calculated. Features are successively eliminated as long as the training set is linearly separable [160, 167]. Instead of the weight vector components w_i another criterion g_i was also used.

$$g_i \; = \; w_i \sum_j x_{ij} \tag{105}$$

The summation is taken over all patterns or only over those which were used for weight vector corrections during the training. Features with smallest values for g_i are eliminated because they have the smallest average contribution to the classification (or to the calculation of the scalar product) [131, 160].

10.3.2. Feature Selection with the Learning Machine

Several feature selection algorithms have been included in the learning machine (Chapter 2.2).

The sign-comparison method [44, 121, 167, 217] trains separately two individual weight vectors with the same training set but with different initial weight vectors. The components of one initial weight vector are set to +1 and of the other weight vector to -1. Components in the two final weight vectors which differ in sign are considered to be less important for the classification. The corresponding features are eliminated from all patterns and the training is repeated. Features are eliminated until the training does not converge.

The dead zone training (Chapter 2.2.7) has also been used for feature selection. Those features are eliminated which do not affect the maximum width of the dead zone significantly [233].

Lidell and Jurs [168] described a modification of the learning machine ("fractional training") for the purpose of feature selection.

The variance-feature-selection [287, 288, 360] computes first a series of weight vectors. The relative variance for all components of the weight vectors are computed and features corresponding to the greatest variances are eliminated. This process is repeated until no linear separability is achieved. This feature selection method is only successful for originally linear separable data. An approximation of the number of intrinsic features in a data set was possible. Mathematical background and valuable hints of an implementation of this method have been given by Zander, Stuper and Jurs [360].

10.3.3. Other Methods

A strategy that seems feasible if the number of patterns is lower than three times the number of dimensions uses cluster analysis of the features [189]. For this purpose, the data set is "inverted": features are treated as objects and the objects are treated as features. A cluster analysis (Chapter 7) shows whether there are groups of features with similar behaviour. One can then take one feature from each group and proceed with other feature selection methods.

The SIMCA method (Chapter 6.3) can be used to delete features which have low modelling and discriminating power [81, 189].

The KNN-method (Chapter 3) [228] and the progressive filter network (Chapter 5.5) [170] have been proposed for feature selection.

If only a small number of features is retained after application of feature selection, then a trial-and-error method may be useful for the next step [297]. Individual features are preliminarily and randomly deleted. A feature is definitively deleted if thereby the classification accuracy is improved.

10.4. Number of Intrinsic Dimensions and Number of Patterns (n/d - Problem)

The main reason for feature selection in many chemical applications of pattern recognition is the necessity to reduce the number of dimensions (d) in relation to the number of patterns (n). If the ratio n/d is less than 3 the chance for a randomly positioned decision plane is not negligible. Because of the importance of the n/d-problem some theoretical considerations are given here.

Suppose a well distributed set of n points in a d-dimensional space (where no subset of d+1 points lies on a (d-1)-dimensional plane). The total number of dichotomies (number of different partitions of these points into two groups) of this data set is 2^n. E.g. for four pattern points there are 16 possibilities to arrange the points into two classes (Table 6). Figure 47 shows a 2-dimensional example for this set of four points; only 2 of the 16 dichotomies cannot be separated by a linear classifier (dichotomy 7 and 10 with patterns a,c in one class and b,d in the other class). All other dichotomies are linearly separable neglecting any physical meaning of the separation. The fraction f

$$f = \frac{\text{number of linearly separable dichotomies}}{\text{number of possible dichotomies}}$$

gives the probability for a linear separability if an arbitrary assignment of 2 classes to all pattern points is made. Therefore, the value of f must be examined if a linear separability was found in a classification problem. Only small values of f give a chance that the linear classifier actually reflects similarities of patterns.

Sophisticated mathematical considerations [391, 396] define f as a function of n and d.

$$f(n,d) = \frac{1}{2^{n-1}} \sum_{i=0}^{d} \binom{n-1}{i} \qquad \text{for} \quad n > d+1$$

$$f(n,d) = 1 \qquad \text{for} \quad n \leq d+1$$

(106)

Figure 48 shows f as a function of the ratio n/d. For n/d < 3 exists a non-negligible probability for a linear separability of random partitioning into two classes. Therefore, a ratio n/d > 3 must be requested for the development of binary classifiers.

TABLE 6. Four patterns a, b, c, d can be placed into two classes in 2^4 different ways. Figure 47 shows for a 2-dimensional example that only two dichotomies (no. 7 and 10) are not linearly separable.

no. of dichotomy	class 1	class 2
1	abcd	–
2	abc	d
3	abd	c
4	acd	b
5	bcd	a
6	ab	cd
--> 7	ac	bd
8	ad	bc
9	bc	ad
--> 10	bd	ac
11	cd	ab
12	a	bcd
13	b	acd
14	c	abd
15	d	abc
16	–	abcd

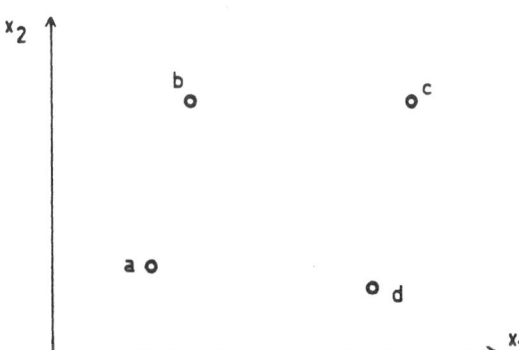

FIGURE 47. Four patterns a, b, c, d in a 2-dimensional pattern space. The patterns points are well distributed because no subset of 3 points lies on a straight line. All arbitrary partitions into two classes are linearly separable except ac/bd and bd/ac (Table 6).

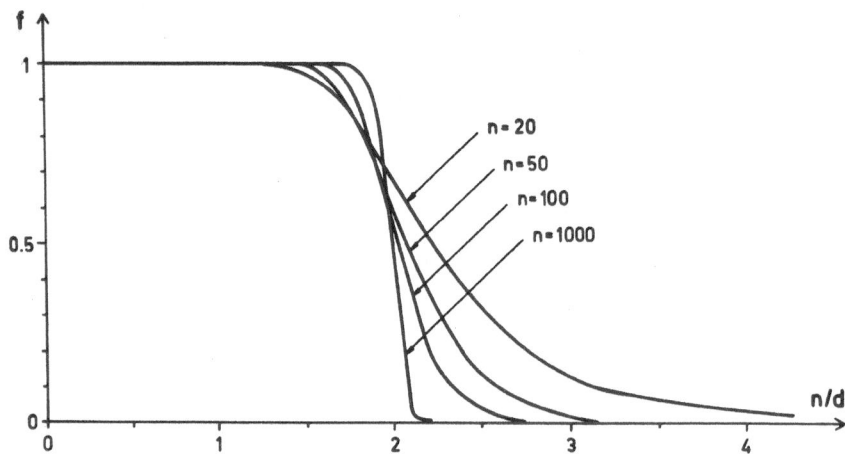

FIGURE 48. Probability f of linear separability of two randomly selected classes; n is the number of patterns and d the number of dimensions.

Gray [94] proposed the elimination of features from the patterns until a predefined value (0.05 or 0.01) for f is reached.

Two problems arise when equation (106) is used to evaluate the significance of a linear classifier.

The first problem concerns the number of dimensions because the number of intrinsic dimensions must be used [245, 380]. Most data sets for chemical classification problems contain correlations between the features. The number of features is therefore often much larger than the minimum number of orthogonal features that will describe the data. The number of intrinsic features can in principle be estimated by factor analysis or other methods (Chapter 10.2.5).

The second problem concerns the a priori probabilities of both classes. These probabilities are predetermined by the composition of the training set. However, equation (106) implies all possible class sizes - ranging from zero to n patterns - with equal frequency.

Both uncertainities in the calculation of f may explain that many valuable classifiers for chemical problems have been calculated from data sets with ratios n/d only less than 3.

Some empirical tests about a random linear separability were made for mass spectra. A set of 250 mass spectral patterns (each with 138

features) have been randomly assigned to two classes. A learning machine
did not converge within reasonable time if one of the classes contained
more than 10 % of the spectra. However, if the classes were built according to chemical classes, the learning machine converged satisfactorly
[310]. Other systematic studies on this topic have been reported by
Rotter and Varmuza [251] for the calculation of binary classifiers by
linear regression. The chance of achieving a satisfactory decision plane
for random classes increases if the population of the classes tends to
be less uniform and if the ratio n/d decreases. Even n/d = 5 was inade-
quate if one class was only weakly (5 %) represented. Similar considerations have been presented by Gray [94].

Extensive experiments about the behaviour of learning machines for
various n/d-ratios have been made by Anderson and Isenhour [8]. They
used Gaussian distributed data with various overlaps of the classes and
they used mass spectra. It was for example shown that linear separability
is very difficult to judge for data with n/d below 2.5 (two classes of
synthetic data with n/d = 2 with nearly complete overlap were found to
be linearly separable in approximately 50 % of all trials).

11. Evaluation of Classifiers

11.1. Principle

The actual merit of a classification method can only be judged in
connection with practical problems. This merit not only depends on the
classification method and problem but also on non-objective criteria
like the demands and knowledges of the user. The basis of a first implementation of a classification method, however, has to consist of objective mathematically defined criteria which characterize the efficiency of
a classifier. The efficiency of a classifier is usually estimated by an
application to a random sample of known patterns (prediction set) which
have not been used for the training. An alternative method is the leave-
-one-out method (Chapter 1.4 [292]).

An evident quality criterion for a classifier seems to be the
percentage of correctly classified patterns (overall predictive ability).
This criterion was used during the first years of pattern recognition
applications in chemistry. However, the overall predictive ability suffers from the fact that it depends extremely on the composition of the
prediction set. If, for example, 90 % of the patterns belong to class 1
and 10 % to class 2, a primitive "classifier" that always predicts
class 1 would have 90 % overall predictive ability. Therefore, overall
predictive ability has to be refused if an objective characterization of
classifiers is necessary.

A recommendable alternative is the characterization of a classifier
by the predictive abilities for all classes separately (Chapter 11.2).
If a classifier should be characterized by a single number one should
refer to a prediction set with equal frequency for all classes. A recommendable quality criterion for this purpose is the transinformation
(Chapter 11.6.2).

Another approach is the comparison of the a priori probability
(before application of a classifier) and the a posteriori probability
(after classification) of the membership of a certain class.

Basic principles of probability theory and information theory are
used in the next Chapters to define objective criteria for the evaluation
of classifiers. All considerations are restricted to binary classifiers
but most concepts can be easily extended to multicategory problems.

Many objections against the usefulness of pattern recognition methods for chemical problems are legitimated because the statistical
evaluation was performed unsatisfactorily in many papers. Further confusions result from a non-uniform terminology (Chapter 11.5). An objective
mathematical evaluation of classifiers is an absolute necessary prerequisite to a further application to actual classification problems.

11.2. Predictive Abilities

A binary classifier is tested by using a random prediction set of
patterns. The composition of the prediction set is given by the
a priori probabilities $p(1)$ and $p(2)$.

$p(1)$: probability of a pattern in the prediction set belonging to
class 1

$p(2)$: probability of a pattern in the prediction set belonging to
class 2

$$p(1) + p(2) = 1 \qquad (107)$$

If a binary classifier is applied to a pattern, two different answers
are possible:

'yes' : pattern is classified into class 1
'no' : pattern is classified into class 2

Therefore, four types of classifications may occur (Table 7).

TABLE 7. Four types of classifications are possible for a binary
classifier.

code	class to which pattern actually belongs	answer of classification	correctness
(1,y)	1	yes	true
(1,n)	1	no	false
(2,y)	2	yes	false
(2,n)	2	no	true

TABLE 8. Probability table for a binary classifier.

p(1,y)	+	p(2,y)	=	p(y)
+		+		+
p(1,n)	+	p(2,n)	=	p(n)
=		=		=
p(1)	+	p(2)	=	1

Each type of classification has its probability $p(m,a)$. These probabilities can be estimated from the classification results obtained with a given prediction set. Table 8 shows relationships between these four probabilities. Two other probabilities are listed in Table 8.

p(y) : probability of a 'yes'-answer
p(n) : probability of a 'no'-answer

$$p(y) \quad + \quad p(n) \quad = \quad 1 \tag{108}$$

Notice that all probabilities in Table 8 depend on the composition of the prediction set. All quality criteria for binary classifiers (with a discrete answer) are derived from this probability table.

Evident criteria for the performance of a classifier are the predictive abilities P_1 and P_2 for class 1 and class 2. The predictive ability of class m is the fraction (or percentage) of correctly classified patterns that belong to class m.

$$P_1 \quad = \quad \frac{p(1,y)}{p(1)} \quad = \quad p(y|1) \tag{109}$$

$$P_2 \quad = \quad \frac{p(2,n)}{p(2)} \quad = \quad p(n|2) \tag{110}$$

p(a|m) denotes a conditional probability (the probability of answer 'a' is calculated only for those events that fulfil the condition 'class m').

Predictive abilities P_1 and P_2 for class 1 and 2 are independent of the composition of the prediction set. P_1 and P_2 together are an objective criterion to characterize a binary classifier. P_1 and P_2 should be specified for any classifier.

In the past the overall predictive ability P was used most frequently to characterize a classifier.

$$P \; = \quad p(1,y) \;\; + \;\; p(2,n) \qquad\qquad \text{or} \qquad\qquad (111)$$

$$P \; = \quad p(1) \; P_1 \;\; + \;\; p(2) \; P_2 \qquad\qquad\qquad (112)$$

The overall predictive ability may be useful for a comparison of different methods within one paper if always the same prediction set is used. But the overall predictive ability is in general not applicable to the comparison of classifiers from different authors because it depends on the composition of the prediction set.

It may be suggested that a classifier is mathematical useful if its overall predictive ability is greater than the overall predictive ability of a "primitive classifier" that always predicts the more frequently occurring class.

$$P \; > \quad \max \; \{p(1), \; p(2)\} \qquad\qquad\qquad (113)$$

A primitive classifier that e.g. always answers 'yes' (class 1 is more frequent) has the predictive abilities $P_1 = 1$ and $P_2 = 0$. A simple example [249] shows that equation (113) cannot judge a classifier correctly. Suppose an actual classifier with $P_1 = 0.90$ and $P_2 = 0.80$. A prediction set with p(1) = 0.60 gives an overall predictive ability P = 0.86 which fulfils equation (113) and the classifier is considered to be useful; however, a test of the same classifier with a prediction set with p(1) = 0.95 gives P = 0.90 and equation (113) is now not fulfilled.

Considerations of a posteriori probabilities and information theory (Chapter 11.4) show that a binary classifier is useful in a mathematical sense if equation (114) is obeyed.

$$P_1 \;\; + \;\; P_2 \;\; > \;\; 1 \qquad\qquad\qquad (114)$$

A classifier with $P_1 + P_2 = 1$ does not change the probabilities of class memberships and does not give information. A classifier with $P_1 + P_2 < 1$ gives reversed answers; a change of the answers 'yes' and 'no' will yield a classifier with $P_1 + P_2 > 1$.

If a classifier has to be characterized by a single number, one should use a prediction set that contains equal numbers of patterns for both classes, t.m. $p(1) = p(2) = 0.5$. In chemical applications such prediction sets are often too small. But one may use an arbitrarily composed predicition set to compute P_1 and P_2. These two criteria can be further applied to derive probabilities $q(m,a)$ for a fictitious random sample that contains equal numbers of patterns in both classes [312]. Consideration of the equalities

$$p(y|1) \quad = \quad q(y|1) \tag{115}$$
$$p(n|2) \quad = \quad q(n|2) \tag{116}$$

gives

$$q(1,y) \quad = \quad P_1 \; / \; 2 \tag{117}$$

$$q(1,n) \quad = \quad (1-P_1) \; / \; 2 \tag{118}$$

$$q(2,y) \quad = \quad (1-P_2) \; / \; 2 \tag{119}$$

$$q(2,n) \quad = \quad P_2 \; / \; 2 \tag{120}$$

Probabilities $q(m,a)$ form a new probability table equivalent to Table 8. From this new probability table, quality criteria can be derived that characterize a classifier by a single number. An evident criterion is the average predictive ability $\bar{P}$

$$\bar{P} \quad = \quad (P_1 + P_2) \; / \; 2 \tag{121}$$

$$\bar{P} \quad = \quad q(1,y) \; + \; q(2,n) \tag{122}$$

A better criterion may be obtained by the application of the information theory (Chapter 11.6).

11.3. Loss

A more differentiated characterization of a classifier is possible if each of the four classification results (Table 7) is attached with a specific loss L(m,a). The total average loss (costs) L of a classifier application is defined by equation (123).

$$L = \sum_{m=1,2} \sum_{a=y,n} p(m,a)\ L(m,a) \qquad (123)$$

Problems with different compositions of prediction sets must be considered in the same way as for predictive abilities (Chapter 11.2).

The results (1,n) and (2,y) with a wrong classification have often different importance. Suppose the classification of a chemical compound whether it is carcinogeneous (class 1) or not (class 2). A wrong classification of an actual carcinogeneous compound (1,n) is much more severe than a wrong classification of a harmless compound (2,y). For such an application, L(1,n) should be set much greater than L(2,y). Optimization of a classifier with respect to the total loss will position the decision boundary in a way that all members of class 1 are actually classified to class 1; assignment to class 2 (harmless compounds) must have a very high accuracy [318, 391].

If the loss for any correct decision is 0 and for any wrong decision 1, the total loss is

$$L = p(2,y) + p(1,n) = 1 - P \qquad (124)$$

11.4. A posteriori Probabilities

A classifier can be interpreted as an instrument that changes the probability that a pattern belongs to a certain class. Before classification there exist a priori probabilities p(1) and p(2) for both classes. A priori probabilities are usually not known explicitly for a given unknown pattern; therefore, equal a priori probabilities should be initially assumed for each class.

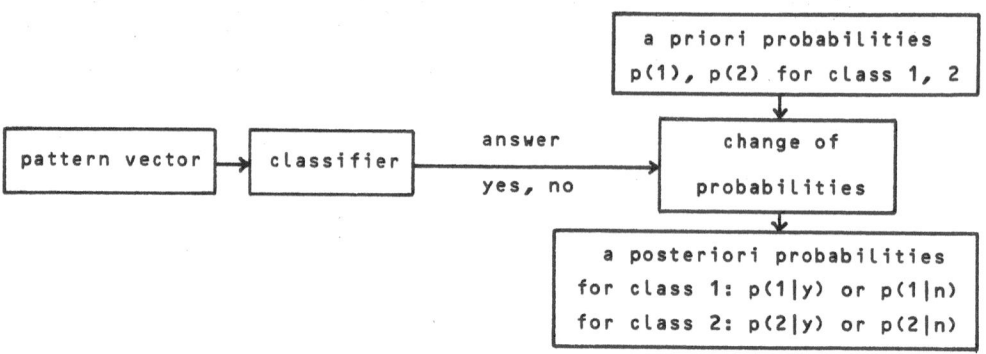

FIGURE 49. A binary classifier changes the probabilities of the membership to class 1 and class 2.

After classification, <u>a posteriori probabilities</u> p(m|a) are known.
The a posteriori probability for a membership to class m depends on the
corresponding a priori probability and on the answer 'a' of the classifier. For instance, the a posteriori probability for class 1 and answer
'yes' is given by equation (125) using Table 8.

$$p(1|y) \quad = \quad \frac{p(1,y)}{p(y)} \tag{125}$$

The a posteriori probability for class 2 and the same answer is

$$p(2|y) \quad = \quad \frac{p(2,y)}{p(y)} \tag{126}$$

with $p(1|y) + p(2|y) = 1$

For the answer 'no' analogous equations exist. Use of the relationships
between the probabilities in Table 8 provides the useful equation (127)
for binary classifiers [248, 249].

$$R \quad = \quad \frac{p(1|a)}{p(2|a)} \quad = \quad Q_a \, \frac{p(1)}{p(2)} \tag{127}$$

answer 'yes': $Q_a = Q_y = P_1 / (1-P_2)$

answer 'no' : $Q_a = Q_n = (1-P_1) / P_2$

Equation (127) gives the ratio R of the probabilities of class 1 to class 2 after classification. R depends on the ratio of the a priori probabilities $p(1)/p(2)$, on the predictive abilities P_1 and P_2 for both classes, and on the classification answer. Figure 50 may help in practical applications of equation (127). Q_y and Q_n may be used for an objective characterization of binary classifiers.

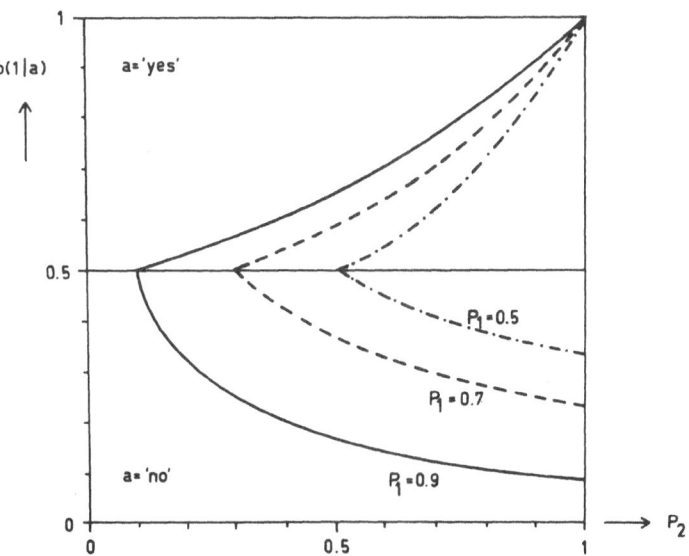

FIGURE 50. A posteriori probability $p(1|a)$ of the membership to class 1 as a function of the predictive abilities P_1 and P_2 and the classification answer 'a' of a binary classifier. Equal a priori probabilities are assumed. The upper half of the diagram is used if the classifier answers 'yes', the lower half for answer 'no'.

Equation (127) can also be applied to successive independent classifications of the same classes. The final ratio of probabilities is independent of the sequence in which the classifiers were applied. Human decisions may be included if estimates about the predictive abilities are known. A hypothetical example for successive classifications is shown in Figure 51.

A posteriori probabilities $p(1|y)$ and $p(2|n)$ give the probability that a given answer of a classifier is true. These values are evidently most interesting for the user of a classifier but it must be emphasized that they are senseless if the a priori probabilities are not defined. Some authors [137, 141, 201, 332] proposed to use p(1|y) and p(2|n) for the evaluation of classifiers. These "confidence values" are calculated from a predicition set and therefore imply the a priori probabilities from this data set. A posteriori probabilities are therefore not suitable for an objective characterization of classifiers.

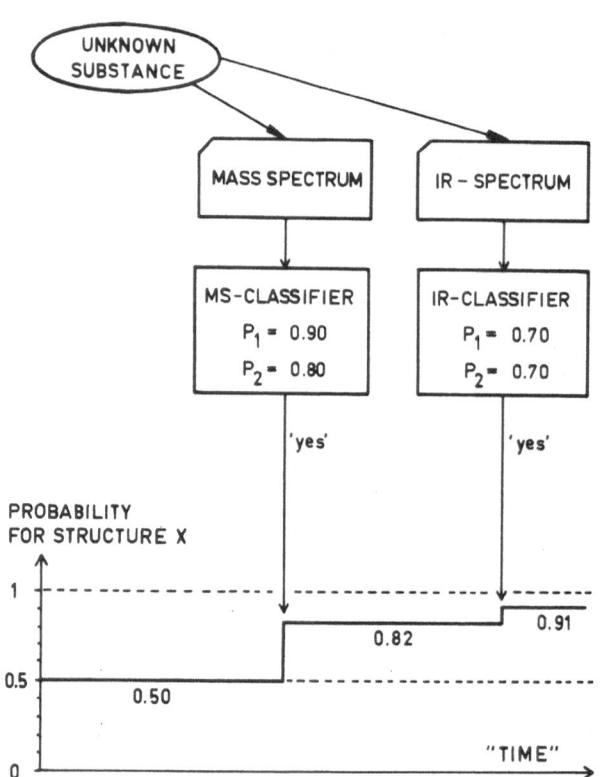

FIGURE 51. Hypothetical example for successive applications of binary classifiers to recognize the presence of a molecular structure X (class 1). The performances of the classifiers are given by their predictive abilities P_1 and P_2. Initially, there are equal probabilities for the presence and absence of X. After application of the mass spectrum classifier the probability for class 1 is 0.82, and the final probability is 0.91 (see equation (127)).

A useful alternative is to relate the a posteriori probabilities to
equal a priori probabilities. Equations (115) to (120) and Table 8 give
[312]

$$p(1|y) \quad = \quad \frac{P_1}{1 + P_1 - P_2} \tag{128}$$

$$\text{for} \quad p(1) \quad = \quad p(2) \quad = \quad 0.5$$

$$p(2|n) \quad = \quad \frac{P_2}{1 - P_1 + P_2} \tag{129}$$

The averaged confidence $\bar{C}$ of the classifier answers is [141]

$$\bar{C} \quad = \quad \{ \, p(1|y) + p(2|n) \, \} \, / \, 2 \tag{130}$$

A weighted a posteriori probability p_W was proposed for the evaluation of classifiers [239].

$$p_W \quad = \quad p(1) \, p(1|y) \quad + \quad p(2) \, p(2|n) \tag{131}$$

Evidently, this criterion depends on the a priori probabilities; therefore, it was recommended to use the minimally possible value as a confidence measure of a classifier. Minimization is done with respect to
a priori probabilities. If the a priori probabilities are not known, this
minimum gives a measure of the most unfavorable case. However, these
criteria suffer from a non-reasonable physical meaning.

11.5. Terminology Problems

Evaluation problems of pattern classifiers for chemical problems
have been treated by several authors. Detailed work about this topic is
reported in about 20 papers listed below alphabetically.

[43]: Chu, Feldmann, Shapiro, Hazard, Geran (1975)
[49]: Clerc, Kutter, Reinhard, Schwarzenbach (1976)
[55]: Coomans, Jonckheer, Massart, Broeckaert, Blockx (1978)
[94]: Gray (1976)
[137]: Kaberline, Wilkins (1978)
[141]: Kent, Gäumann (1975)

[159]: Kowalski, Jurs, Isenhour, Reilley (1969)
[165]: Lam, Wilkins, Brunner, Soltzberg, Kaberline (1976)
[172]: Lowry, Isenhour, Justice, Mc Lafferty, Dayringer, Venkataraghavan (1977)
[201]: Mc Lafferty (1977)
[239]: Richards, Griffiths (1979)
[244]: Ritter, Lowry, Woodruff, Isenhour (1976)
[249]: Rotter, Varmuza (1975)
[279]: Soltzberg, Wilkins, Kaberline, Lam, Brunner (1976)
[280]: Soltzberg, Wilkins, Kaberline, Lam, Brunner (1976)
[310]: Varmuza, Krenmayr (1973)
[312]: Varmuza, Rotter (1976)
[313]: Varmuza, Rotter (1978)
[314]: Varmuza, Rotter (1979)
[332]: Wilkins (1977)
[357]: Woodruff, Snelling, Shelley, Munk (1977)

Unfortunately, not all authors use the same terminology. The following list may prevent possible confusions (the probabilities have been defined in Chapter 11.2).

$p(y|1) = P_1$: predictive ability for class 1,
recall [49, 172, 201],
sensitivity [55].

$p(n|2) = P_2$: predictive ability for class 2,
selectivity [55].

$p(1|y)$: a posteriori probability for class 1,
confidence or reliability of answer 'yes' [141, 201, 310, 312],
precision [49],
percent correct answers [172].

$p(2|n)$: a posteriori probability for class 2,
confidence or reliability of answer 'no' [141, 310, 312].

$p(1,y) + p(2,n) = P$: (overall) predictive ability,
efficiency [55].

$\{p(y|1) + p(n|2) - 1\}100$: Youden-index [55].

11.6. Application of Information Theory

11.6.1. Introduction to Information Theory

Only a very short introduction to some basic ideas of information theory [189, 410, 412, 417, 420, 427, 430] is given here in order to define those terms which are used for the evaluation of classifiers.

Information theory is the mathematical background in the area of communication but can be also used to describe the output of a series of experiments. It deals with probabilities and is therefore not applicable to single experiments.

The information of an event is considered to be only a function of the probability p_i of this event. The importance or content or meaning of an event to a human interpreter is completely neglected in the information theory. The information I_i of an event i is defined according to a proposal by Hartley [419].

$$I_i = -ld\ p_i \qquad (\ ld = {}_2log\ ) \qquad (132)$$

The unit of information is 1 bit. An event has an information of 1 bit if the probability is 0.5 for this event.

Suppose a certain experiment A which may have n possible discrete results (outputs, events). All outputs are mutually exclusive and have the probabilities p_i (i = 1...n).

$$\sum_{i=1}^{n} p_i = 1 \qquad (133)$$

Experiment A can be considered as a source which randomly produces discrete signals. The expected value for the information is given by the weighted average of I_i over all possible outputs.

$$H(A) = -\sum_{i=1}^{n} p_i \cdot ld\ p_i \qquad (134)$$

H(A) is called the entropy of the discrete source A (Shannon [433,434]).

The unit of entropy is 1 bit and equation (134) shows that $H(A) \geq 0$. The "entropy in information theory" is closely related to the "entropy in thermodynamics" [410, 427]. $H(A)$ is a measure of the uncertainty or capacity of the source A. Before experiment A is performed, an uncertainty of $H(A)$ bit exists. After the experiment one knows the result and therefore the uncertainty is zero. On the average, an information of $H(A)$ bit results from this experiment. The amount of information which can be achieved by an experiment depends on the uncertainty (entropy) which exists before the experiment is carried out.

A binary classification problem can be considered as an experiment A with 2 possible results: '1' for class 1 and '2' for class 2. Suppose a large set of patterns with a priori probabilities p(1) and p(2) for both classes. The entropy of this data set with respect to this classification is given by equation (134) and re-formulated in equation (135).

$$H(A)_{binary} = -p(1).ld\ p(1) - \{1-p(1)\}.ld\{1-p(1)\} \tag{135}$$

Figure 52 shows the entropy of a binary discrete source as a function of the probability of class 1.

An ideal classifier which makes no mistakes leaves no uncertainty after classification; $H(A)$ is therefore the maximum information which

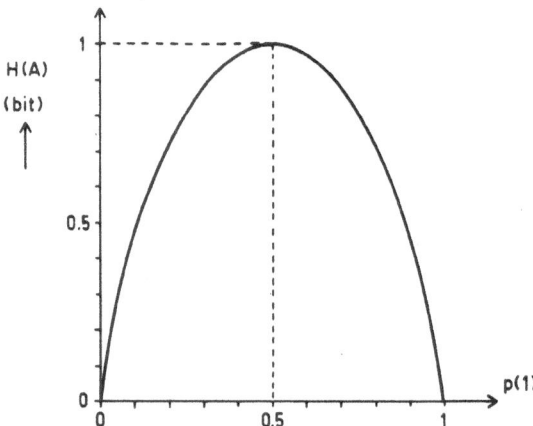

FIGURE 52. Entropy $H(A)$ of a binary discrete source as a function of the probability p(1). $H(A)$ has the maximum value of 1 bit if both results of the source are equally probable.

can be obtained. However, an ideal classifier cannot give much information if the result of the classification is, on an average, only less uncertain (one class is a priori much more frequent than the other). A binary classifier produces the maximum information of 1 bit if no mistakes are made and if equal a priori probabilities are assumed for both classes. Multicategory classifications can be treated in the same way by using equation (134).

11.6.2. Transinformation

Actual classifiers make more or less errors. The initial uncertainty H(A) about the membership to a certain class is therefore not completely eliminated (Figure 53). The difference of the entropies before and after classification is called the transinformation R (syn-entropy, rate of information, information gain). R depends on the previous entropy H(A) and on the quality of the classifier. If the transinformation R is used for an evaluation of classifiers, one should refer to a pattern set with maximum uncertainty before classification (t.m. for a binary problem p(1) = p(2) = 0.5 and H(A) = 1 bit) [312, 313]; the transinformation in this case will be called standard (trans)information R_o. R_o is obtained as an averaged value if a classifier is applied to a random sample of patterns with equal numbers of patterns in each class.

The standard transinformation R_o is not necessarily the maximum amount of information that can be obtained by a given classifier. The maximum amount R_{max} is called channel capacity [417, 427]. R_{max} would be obtained for a random sample of patterns with optimally adjusted a priori probabilities. Because the optimal a priori probabilities depend on the classifier, the channel capacity is not suitable for the evaluation of classifiers. The only useful starting point for the classification of unknown chemical patterns is to assume equal probabilities for all classes.

The standard transinformation R_o is a function of the probabilities defined in Table 8.

$$R_o = \sum_{m=1,2} \sum_{a=y,n} q(m,a) \cdot ld \frac{q(m,a)}{q(m)\, q(a)} \qquad (136)$$

The probabilities in equation (136) are denoted by 'q' (instead of 'p')

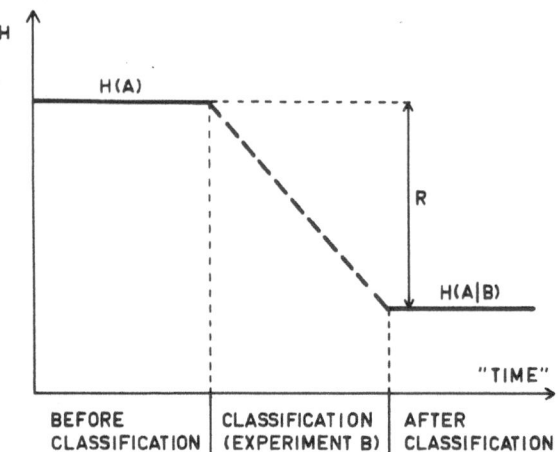

FIGURE 53. A classifier for a binary classification problem A reduces the initial entropy (uncertainty) H(A) to a lower value H(A|B). The transinformation is the difference between the entropies.

to indicate that a (fictitious) pattern set with equal numbers of patterns in each class was used.

The probabilities in equation (136) can be computed from the predictive abilities P_1 and P_2 by using equations (117) to (120). The determination of the standard transinformation from the predictive abilities for both classes is facilitated by the diagram shown in Figure 54.

The transinformation is measured in bit and equation (137) holds for binary classifiers.

$$0 \leq R \leq 1 \tag{137}$$

R = 0 means that there is no connection between classification and actual class membership. It can also be shown that all classifiers with $P_1 + P_2 = 1$ (Chapter 11.2) provide zero transinformation.

R = 1 means that all classifications are correct (or that all classifications are wrong !).

The standard transinformation R_0 can be advantageously used as a single number to characterize the performance of classifiers. Classifiers are distinguishable by R_0 even if they have identical values for the average predictive ability $\bar{P}$ as shown in Table 9.

TABLE 9. Comparison of two binary classifiers. P_1, P_2: predictive abilities for both classes, $\bar{P}$: average predictive ability, R_o: standard transinformation. The classifiers are distinguishable by R_o but not by $\bar{P}$.

Classifier	P_1	P_2	$\bar{P}$	R_o (bit)
1	0.7	0.9	0.8	0.296
2	0.8	0.8	0.8	0.232

The concept of classifier evaluation by the standard transinformation suffers mainly from the complexity of equation (136) and from less intuitivness. On the other hand, the advantages are: only a single number is used for the characterization of a classifier and R_o is a theoretically founded criterion which is independent of the composition of the prediction set. The same concept can be used for classifiers with a continuous response (Chapter 11.7) or multicategory problems.

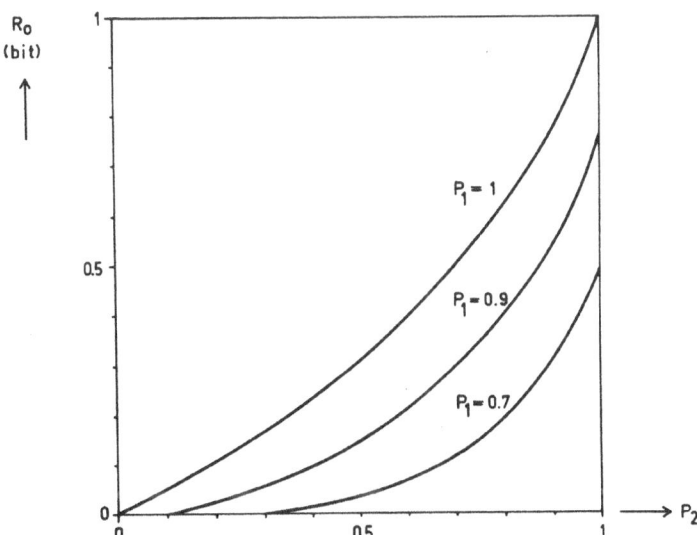

FIGURE 54. Standard transinformation R_o for binary classifiers as a function of predictive abilities P_1 and P_2 for both classes. Equal a priori probabilities for both classes are assumed [313].

Remarks on equation (136) concerning transinformation :

The entropy of the classification problem (experiment A) is given by equation (134) and is re-formulated by equation (138).

$$H(A) = - \sum_m p(m) \cdot ld\ p(m) \tag{138}$$

p(m) : a priori probability of class m

The classifier gives discrete answers. After classification (experiment B) a conditional entropy H(A|B) remains. The transinformation R is

$$R = H(A) - H(A|B) \tag{139}$$

The conditional entropy H(A|B) is calculated analogously to equation (134) but with the conditional probabilities p(a|m) instead of p_i. These probabilities are estimated for all answers 'a' and classes 'm' from a prediction set (Chapter 11.2). The entropy which is connected with a certain answer - e.g. answer 'yes' - is given in equation (140).

$$H(A|a=yes) = - \sum_m p(m|a=yes) \cdot ld\ p(m|a=yes) \tag{140}$$

The weighted average over all possible answers 'a' is

$$H(A|B) = - \sum_a \left[p(a) \cdot \sum_m p(m|a) \cdot ld\ p(m|a) \right] \tag{141}$$

Insertion of equations (141) and (138) into equation (139) and use of the relationships given in Table 8 yields, after some trivial but laborious mathematical transformations, the transinformation R for a binary classifier.

$$R = \sum_{m=1,2} \sum_{a=y,n} p(m,a) \cdot ld\ \frac{p(m,a)}{p(m)\ p(a)} \tag{142}$$

Equation (142) is equivalent to equation (136) but if the probabilities p are used, the transinformation depends on the a priori probabilities of the prediction set (Figure 55).

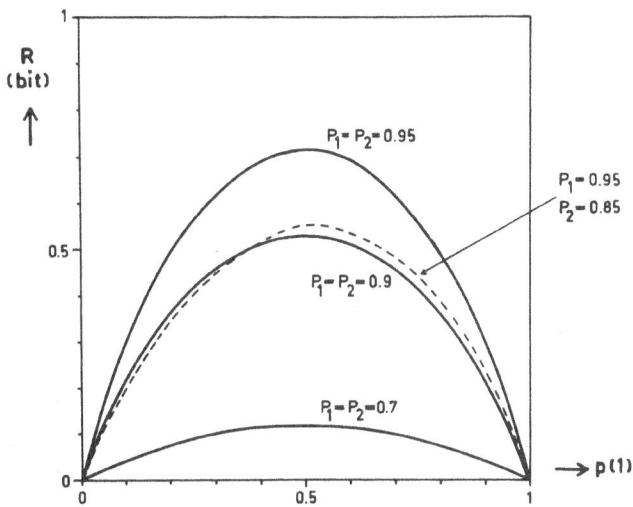

FIGURE 55. Transinformation R for a binary classifier as a function of the a priori probability p(1) of class 1 for four different classifiers with various predictive abilities P_1 and P_2.

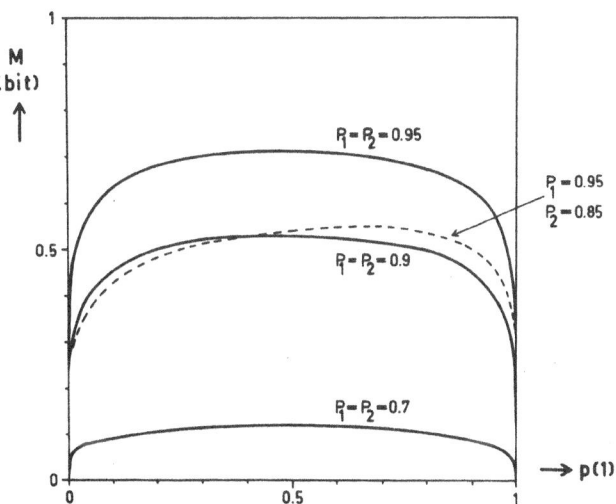

FIGURE 56. Figure of merit M for a binary classifier as a function of the a priori probability p(1) of class 1 for four different classifiers with various predictive abilities P_1 and P_2.

11.6.3. Figure of Merit

The transinformation R (equation (142)) is not suitable for the characterization of a classifier because R depends on the composition of the prediction set (Figure 55). In order to diminish this undesirable effect, Wilkins et.al. [137, 165, 279, 332] normalized the transinformation R relative to the initial entropy H(A) of the prediction set.

$$M \quad = \quad R \, / \, H(A) \tag{143}$$

M is called figure of merit (or relative transinformation). Figure 56 shows that the figure of merit still depends on the a priori probabilities of the predicition set but may give acceptable results if the number of patterns in both classes does not differ significantly.

11.7. Evaluation of Classifiers with Continuous Response

Classifiers with a continuous response have been described in Chapter 2.6.1. Evaluation of this type of classifiers requires the characterization of the overlap region of the probability density functions (Figure 57).

The Bhattacharyya coefficient B is a measure of the overlap of two density functions.

$$B \quad = \quad \int_{-\infty}^{+\infty} \{g_1(s) \cdot g_2(s)\}^{1/2} \, ds \tag{144}$$

 B = 0 : no overlap
 B = 1 : identical density functions

Another approach to the characterization of a classifier with a continuous response is the transinformation (Chapter 11.6.2). The entropy of a continuous distribution of signals depends on the unit in which the signal is measured and is therefore less useful. In contrast to the entropy the transinformation is independent of the unit of measure [417, 427]. A binary classifier with a continuous response s corresponds to a communication channel with two discrete inputs (patterns of class 1 or class 2) and one continuous output s.

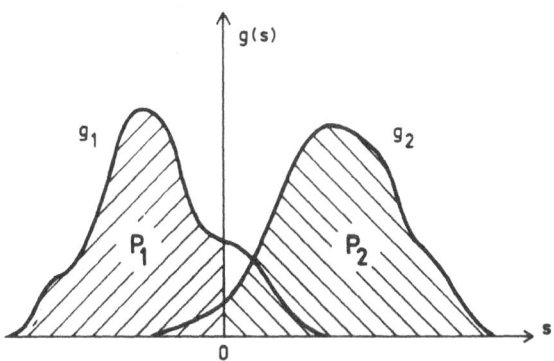

FIGURE 57. Probability density functions for a binary classifier with a continuous response s; g_1 and g_2 are the probability densities for class 1 and 2. The shaded areas correspond to the predictive abilities of a discrete classifier with a zero threshold.

The a priori probabilities p(1) and p(2) of both classes at the input are connected by the classification procedure with the probability densities $g_1(s)$ and $g_2(s)$ at the output. The transinformation R_c of this channel is given by equation (145).

$$R_c = \int_{-\infty}^{+\infty} \left[\sum_{i=1}^{2} p(i) \cdot g_i \cdot ld\, \frac{g_i}{g} \right] ds \qquad (145)$$

with $\quad g = p(1)\, g_1 + p(2)\, g_2$

$\qquad g_i = g_i(s)$

R_c depends on the a priori probabilities. In a similar way, as described in Chapter 11.6.2, one should characterize a classifier with a continuous response by the standard transinformation $R_{c,o}$ which is obtained when both classes are equally probable [314]. Insertion of p(1) = p(2) = 0.5 into equation (145) yields $R_{c,o}$.

$$R_{c,o} = 0.5 \int_{-\infty}^{+\infty} \left[g_1\, ld\, \frac{2\, g_1}{g_1 + g_2} + g_2\, ld\, \frac{2\, g_2}{g_1 + g_2} \right] ds \qquad (146)$$

$$0 \leq R_{c,o} \leq 1$$

$R_{c,o}$ = 0 bit : identical density curves, $g_1 = g_2$ for all values of s

$R_{c,o}$ = 1 bit : no overlap of the density curves

$R_{c,o}$ is an objective measure of the quality of a binary classifier with a continuous response and is independent of the composition of the prediction set. $R_{c,o}$ can be directly compared with the standard transinformation R_o of a classifier with a discrete response. Calculation of $R_{c,o}$ only requires the probability densities g_1 and g_2.

11.8. Confidence of Predictive Abilities

The predictive abilities of each class are fundamental and easy to calculate criteria of a classifier. Therefore, it is important to estimate the confidence of these values.

A frequently used experimental method adds noise or other random or systematic errors to the patterns of the prediction set. The variations of the predictive abilities are calculated [123, 200, 270, 297, 310, 327].

Mc Gill and Kowalski [200] proposed a diagram with predictive ability versus recognition ability to show the influence of noise and to compare classification methods.

Another experimental approach is to alter the composition and/or size of the training set. The variations of the predictive abilities show the "stability" of the method [94, 123, 160].

Numerical experiments with random patterns or with chemical patterns but random classes are sometimes used to mark the lowest limit for a classifier performance [94, 251, 288, 303, 310].

An extensive mathematical treatment about the confidence of predictive abilities was given by Highleyman [384]. Let P be the estimated predictive ability that was obtained from a prediction set with n patterns. The confidence interval $P_{low} \ldots P_{high}$ is a function of P, n, and the probability that the true value lies within the interval. An approximative calculation of the confidence interval for a statistical security of 95 % can be carried out by use of equation (147) [384].

$$P_{high,low} = P \pm 1.96 \; \frac{P \; (1-P)}{n} \qquad (147)$$

Figure 58 was calculated without approximations. An example may show the use of this diagram. Suppose a predictive ability of P = 0.90 was found for a prediction set with n = 100 patterns. Figure 58 gives P_{low} = 0.82 and P_{high} = 0.95; equation (147) yields approximative values of 0.84 and 0.96. The actual predictive ability lies between these values with a probability of 95 %.
Thus, another classifier with a predictive ability of 0.93 cannot be considered to be better with statistical significance. Confidence inter-vals of predictive abilities have hitherto been considered only rarely in chemical applications of pattern recognition [94, 248].

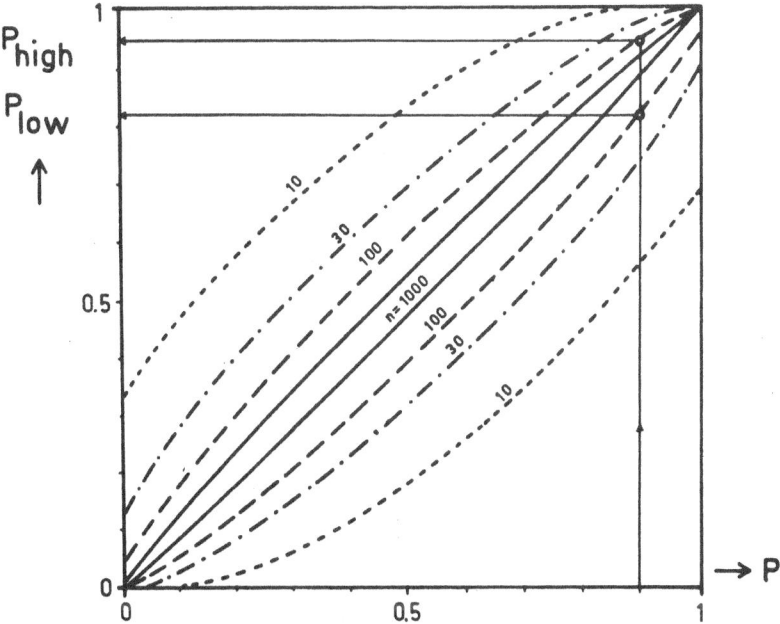

FIGURE 58. Confidence intervals of predictive abilities for a statisti-cal security of 95 %. P: calculated predictive ability, n: number of patterns in the prediction set, P_{low}, P_{high}: boundaries of the confi-dence interval [248, 384].

11.9. Comparison with the Capability of Chemists

Direct comparisons of pattern recognition methods with the capability of chemists have been reported only in a few papers. Such comparisons must be necessarily restricted to a small number of examples because it is hard to find chemists who classify e.g. some hundred spectra into a couple of classes.

Crawford and Morrison [60] found for a sophisticated mass spectral interpretation program a capability of the same order as that for an undergraduate student. A similar result has been reported [119] about the interpretation of binary encoded infrared spectra. Kowalski et. al. [162] emphasized the superiority of pattern recognition methods in the interpretation of multidimensional data.

One should never forget that the aim of pattern recognition methods in chemistry is not the substitution of the chemist. Automatic classification methods can only be a tool to make the interpretation of data easier or faster or to present new previously overseen hints. Therefore, another comparison seems to be more useful: chemist on one side and chemist plus pattern recognition on the other side.

Part B

Application of Pattern Recognition Methods in Chemistry

12. General Aspects of Pattern Recognition in Chemistry

Pattern recognition methods are automatic procedures for classifying observed individuals into discrete groups on the basis of a multivariate data matrix [264]. Several hundred papers published in the last ten years deal with chemical applications of pattern recognition methods. It has been shown by more than 300 authors that pattern recognition can be used for chemical classification problems for which no direct or theoretical approaches exist. However, one must confirm that only a few papers show a really impressive success or a promising utility of pattern recognition methods for important chemical problems. Many papers report about a first (or new) application in chemistry of well-known (or slightly modified) pattern recognition methods. It seems unavoidable in an interdisciplinary field that sometimes methods are improperly used. After the first years of too optimistic enthusiasm on one side and pessimistic refusal on the other side, one can notice some consolidation and peace in the last years.

Wonders must not be expected from conceptually rather simple methods of pattern recognition for the solution of very complex problems (like the interpretation of a spectrum or the description of structure-acti-vity-relationships). Pattern recognition can be seen only as one part of a computer-assisted interpretation system for chemical data. Accentuation should be given to "assisted" because a complete processing by a machine of sophisticated data interpretations in chemistry is unrealistic and uneconomical at least for the next two decades.

From the point of view of system theory an analytical chemical system consists of three parts [316].
1. Sign maker: interactions between a sample and a reagent produce measurements.
2. Sign interpreter: chemical information is extracted from the measurements by pattern recognition methods in a general sense.
3. Control unit: for sign maker and interpreter.

Intensive applications of pattern recognition methods in chemistry were started with pioneering works about spectral interpretations by Isenhour, Jurs, and Kowalski (1969 - 1971). Numerous papers deal with the automatic prediction of molecular structures from mass spectra, infrared spectra and nuclear magnetic resonance spectra (Chapter 13). Predictive abilities of 80 to 95 % are typical of these applications.

Comparison of unknown spectra with library spectra (library search) is
usually not ascribed to pattern recognition. Pattern recognition methods
have been successfully used in <u>chromatography</u> to describe the similarity
of stationary phases (Chapter 14). <u>Classification of materials</u> is a promising field for pattern recognition techniques (Chapter 16). Assignment
of archaeological artefacts according to the content of trace elements is
a typical example. Similar problems are the classification of technolo-
gical materials, food, bacteria, and plants. Perhaps the most valuable
application of pattern recognition methods in chemistry will be the
<u>prediction of biological activities</u> of compounds from their molecular
structures (Chapter 17). Pattern recognition is used as a part of computer-assisted medical diagnosis by using experimental data from <u>clinical</u>
<u>chemistry</u> (Chapter 18). Identification of oil spills and classification
of aerosols are applications of pattern recognition in <u>environmental</u>
<u>chemistry</u> (Chapter 19). It was also proposed to describe and compare
analytical methods by pattern recognition techniques (Chapter 20).
<u>Recognition of</u> printed and handwritten <u>characters</u> may find more use in
chemistry for a comfortable processing and coding of chemical structures [358].

Several <u>packages of computer programs</u> have been developed for chemical applications of pattern recognition. The "RECOG"-system was developed at the Lawrence Livermore Laboratory, California [57]. Kowalski
et. al. [78, 146] made available their collection of computer programs
"ARTHUR" to numerous laboratories. These programs contain all important
pattern recognition methods [179, 180]. Implementation and use of this
package, however, requires a profound knowledge of programming and
pattern recognition. An improved version of "ARTHUR" considers measurement uncertainties [101]. Wilkins [332] reported an interactive pattern
recognition system for spectral interpretation and a semiautomated structure analysis. Jurs et. al. [291] developed an interactive, modular program system "ADAPT" for the investigation of structure-activity-relationships. A package of pattern recognition programs have been implemented
on a dedicated mini-computer that is usually supplied with energy-disper-
sive X-ray spectrometers [326].

Sjöstrom and Kowalski [267] reported an extensive comparison of
five pattern recognition methods and several preprocessing methods
(mainly taken from the "ARTHUR"-package). Six real data bases previously described in the literature were investigated and the advantages and
limitations of the methods discussed. The authors emphasized that methods
which can deal with outliers are desirable in many chemical applications.

Beginners in pattern recognition should not be frightened by program systems with a length of some mega-bytes as described in the literature. Only a basic knowledge of programming is necessary to write some short and simple programs for own classification problems (e.g. for the classification by distance measurement, learning machine, KNN-method). No problem at all is the use of already computed classifiers.

The actual value of pattern recognition methods in chemistry seems to be uncertain at the moment [47, 363]. The future will show whether mathematical pattern recognition methods are useful only for some specialized problems in chemistry or whether they represent a powerful and advantageous problem solving tool for a variety of data interpretations. Three points will require more attention than in the past to clarify the situation.

1. More cooperation between professional chemists (who know the problems) and professional "pattern recognizers" (who know the methods).
2. More disposal of packages with classification methods (or classifiers) for a wide use by many chemists that need not be experts in pattern recognition or computer programming.
3. More basic understanding by more chemists what pattern recognition is and more readiness by some chemists to let a machine do jobs that might be related to something like "intelligence".

Pattern recognition is an approach to data reduction which is essential today because "we already have far more chemical information than anyone knows to manage" [150].

13. Spectral Analysis

13.1. Mass Spectrometry

13.1.1. Survey

Interpretation of low resolution mass spectra is the field with the greatest number of applications of pattern recognition techniques in chemistry. Numerous methods of preprocessing, feature selection, training, and evaluation have been tested with mass spectral data in about 100 papers. Probably the first application of pattern recognition ideas in mass spectrometry has been reported by Raznikov and Talroze [235]; this Russian paper is summarized in [224].

Classification problems with low resolution mass spectra have become a standard in chemometrics for testing new pattern recognition ideas. However, comparisons between the results reported in these papers are difficult or impossible because the data bases used were different and evaluation of the classifiers was performed in different ways.

Preference of mass spectral data for pattern recognition work has several reasons [170]: computer-readable files were early available, peak heights at integral mass numbers can easily be used as pattern components, mass spectral data contain large amounts of information related to molecular structure, the theory of mass spectral fragmentation is incomplete, and large numbers of spectra in gas chromatography - mass spectrometry analysis require an automatic interpretation. Most of the papers deal with the recognition of molecular structures from low resolution mass spectra. The aim of pattern recognition classifiers is not a complete identification of an unknown but rather a preliminary classification into defined groups of chemical compounds. In several papers mass spectral data only served as a vehicle to introduce new pattern recognition techniques into chemometrics.

Although a lot of effort has been devoted to the development of pattern recognition in chemistry, exciting practical results in mass spectrometry are rather meagre. The main reasons seem to be the chronic use of data sets which are too small, and some ignorance of basic principles in statistics in the earlier papers [50]. It was questioned by some

authors whether pattern classifiers work satisfactorly in mass spectrometry to solve more than trivial problems [51, 141, 203, 217, 318].
Adams and Sedgwick [3] reported a reciprocal correlation between predictive ability for mass spectral classifications and the molecular complexity (characterized by the number of characters in the Wiswesser nota-
tion). Crawford and Morrison [60] stated for their computerized mass
spectral interpretation system which includes some pattern recognition
techniques: "Its capability is of the same order as that of an undergraduate student who had completed a course of 16 lectures on organic mass
spectrometry".

Due to the large number of papers dealing with pattern recognition
applications in mass spectrometry it may be useful to list some reviews
on this topic (alphabetically).

 [11]: Bailey (1979)
 [39]: Chapman (1978)
[115]: Isenhour, Jurs (1971)
[117]: Isenhour, Jurs (1973)
[128]: Jurs, Isenhour (1975)
[148]: Kowalski (1974)
[200]: Mc Gill, Kowalski (1978)
[205]: Meisel, Jolley, Heller, Milne (1979)
[206]: Mellon (1975)
[207]: Mellon (1977)
[208]: Mellon (1979)
[224]: Pesyna (1975)
[323]: Ward (1971)
[324]: Ward (1973)

Most applications of pattern recognition methods in mass spectrometry deal with sets of some hundred compounds from very different chemical classes. In Table 10 are listed some works that have been done with
more specialized data sets.

13.1.2. Representation of Mass Spectra as Pattern Vectors

A great variety of methods have been used for the representation
of mass spectra as pattern vectors, for preprocessing and feature selection. An evident method for the representation of a low resolution
mass spectrum in a multidimensional space is to use directly the original
mass spectrum as a vector. Peak heights x_i at integral mass numbers i
correspond to the vector components. Elimination of very small peaks
(e.g. lower than 0.5 or 1 % of the base peak) is useful [1, 165, 193].

TABLE 10. Pattern recognition applications in mass spectrometry with specialized data sets.

Classification problem	Literature
Discrimination between alkanes and alkenes	[235, 258]
Discrimination between monocyclic naphthenes and monoolefins	[142]
Molecular structure of alkylbenzenes	[193]
Recognition of polycyclic aromatic hydrocarbons	[49]
Cluster analysis and factor analysis of sulfur-containing compounds	[105, 106, 143, 268]
Recognition of phosphonate insecticides	[192]
Molecular structure of steroids	[248, 250, 251, 314, 315]
Discrimination between stereoisomers of N-acetylhexosamines	[317]
Sequence analysis of di- and oligo-deoxyribonucleotides	[33, 34, 330]

In order to reduce the dynamic range and to improve the classification result the square root or the logarithm of x_i are frequently used. New logarithmic features x_i' which lie in the intensity range 0 to 100 can be obtained by equation (148) [250].

$$x_i' = (100 \log x_i + 100) / 3 \qquad \text{if } x_i > 0.1 \text{ \% base peak}$$

$$x_i' = 0 \qquad \text{if } x_i \leq 0.1 \text{ \% base peak}$$

(148)

In binary encoded mass spectra [418] features are '1' if peak heights are greater than a threshold (e.g. 1 % base peak); otherwise, they are '0'. This drastic data reduction often gives remarkably good results which confirm the fact that the mass number is much more important than the peak height. The division of the peak height range into 3 to 8 intervals seems to be very appropriate for pattern recognition applications.

From the information theory (Chapter 11.6.1) it is known that a signal contains most information if all possible values of the signal have
equal probability. To generate pattern vectors with equally distributed
features, intensity levels must be created for each mass number. Each
level should have the same probability in the spectral file. If there
are k intensity levels and d mass numbers, a set of (k-1)'d threshold
values must be stored for spectral preprocessing [250].

Normalization of the features is performed with respect to the sum
of all peak heights ("total ion current") [141], or with respect to the
sum of the squared peak heights (t.m. all spectra lie on a hypersphere)
[59, 235]. A valuable preprocessing is "normalization to local ion
current" [217, 250, 413]. Each peak height is divided by the sum of peak
heights in a window of several mass units.

$$x_i' \;=\; x_i \left/ \sum_{m=i-\Delta i}^{i+\Delta i} x_m \right. \tag{149}$$

Window widths ($2\Delta i + 1$) between 3 and 7 mass units gave good results.
This procedure enlarges isolated peaks or peak groups.

New features were generated from mass spectra by various methods.
Instead of the mass number the difference to the molecular ion was used
[1, 217]. Peak heights at 2 to 16 adjacent mass numbers were added to
new features [160, 192]. Ratios of peak heights were found to be good
discriminators in some classification problems [33, 34, 317, 318].

Ion series spectra are obtained by summing up peaks with a mass
difference of 14 ("modulo 14 summed"). Applications of these pattern
vectors containing 14 components were investigated by several authors
[12, 59, 143, 144, 270, 271, 283].

More complicated preprocessing methods involve calculation of cross
terms [124, 160] or moments [16, 17, 18], Fourier transformation [122,
319, 321] or the "linear CNDF method" [134, 135].

It is preferable to apply mass spectrometric knowledge during the
generation and selection of features. This concept was used for the
classification of phosphonate insecticides [192], sulfur compounds [105,
106], polycyclic aromatic hydrocarbons [49], and other compounds [75, 95,
202, 229, 281].

Data reduction methods which are widely used in library search
have also been tested for pattern recognition. The k largest peaks in a
spectrum (k = 5 to 50) , or selection of the k largest peaks in mass
intervals of length l (k = 1 to 2, l = 7 or 14) were utilized for

generation of pattern vectors [257, 258]. A selection of a subset of
mass numbers was performed by considering the number of peaks [165], or
the sum of the peak heights at each mass number [251, 300].

A different method of selecting the most important peaks in a mass
spectrum was proposed by Nägeli and Clerc [217]. Figure 59 shows the procedure to eliminate "unimportant" peaks in order to generate "significant
mass spectra". In binary significant spectra, peak heights of all significant peaks are set to value '1'. Significant mass spectra with normali-

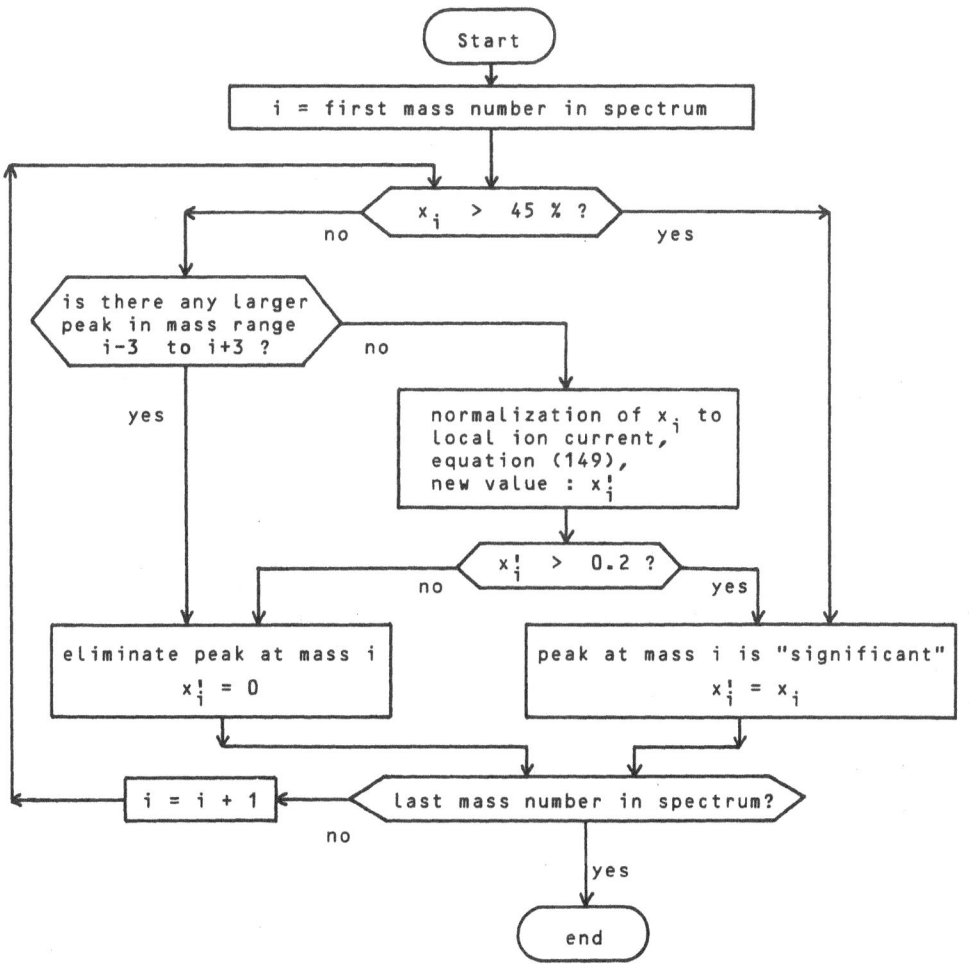

FIGURE 59. Scheme for generating significant mass spectra (with peak
heights x_i') from normal mass spectra (peak heights x_i in % base peak).

zation to local ion current contain only significant peaks according to Figure 59 but peak heights are taken from the full spectrum which has been normalized to the local ion current [250].

In general, mass spectra contain a great amount of redundant information and therefore a drastic data reduction is possible without a significant decrease of the classification results.

Extensive comparisons between different representations of mass spectra have been reported [200, 250, 251, 318].

Table 11 compares some preprocessing methods for mass spectra of steroids [251]. Normal spectra and spectra normalized to local ion current show the best classification. These results are valid only for a certain classification procedure and only for steroid mass spectra. Nevertheless, qualitative conclusions may be drawn also for other methods or other types of compounds.

13.1.3. Determination of Molecular Formulas and Molecular Weights

One of the first pattern recognition applications in mass spectrometry was the attempt to determine the molecular formula by a decision tree [120, 128, 129]. The decision tree contained several binary classifiers. Each of the classifiers decided whether a compound contains more atoms than a given number. A run through the decision tree yields the molecular formula of an unknown whose low resolution mass spectrum is known. A tree with 26 classifiers was necessary for a set of 346 compounds of formulas $C_{1-7}H_{1-16}O_{0-3}N_{0-2}$. For mass spectra with an artificial noise on peak heights (standard deviation 1 to 20 %) 96 to 100 % of the formulas were determined correctly. However, the increased total risk for consecutive classifications and the limited range of compounds will probably prevent practical applications of this method (Chapter 2.6.3).

Recognition of the number of carbon atoms, hydrogen atoms, oxygen atoms, and nitrogen atoms was often used to test several classification methods, e.g. learning machine [85, 121, 122, 131], learning machine with a dead zone [123], KNN-method [158], least-squared error method [160, 229], feature selection methods [121, 158, 229], and other methods [16, 17, 18]. Typical predictive abilities lie between 80 and 95 %.

A detailed investigation with 539 compounds of formulas $C_{6-8}H_nO_{0-2}$ was reported by Mc Gill and Kowalski [200]. Several pattern recognition methods have been applied to separate compounds containing no oxygen, one oxygen atom and two oxygen atoms. The best results were obtained with

TABLE 11. Comparison of preprocessing methods for mass spectra [251].
Training set: 262 steroid mass spectra, 75 mass numbers selected by the
maximum Fisher ratio. Training: linear regression. For each preprocessing
method a set of 14 binary classifiers has been trained that recognize
molecular structures in a steroid molecule. Prediction set: 262 steroid
mass spectra differing from the training set spectra. $\bar{P}_1$ and $\bar{P}_2$ are the
predictive abilities for class 1 and class 2, averaged for all 14 molecular structures, $\bar{I}_0$ is the corresponding averaged standard transinformation.

preprocessing	predictive ability		transinformation
	$\bar{P}_1$ (%)	$\bar{P}_2$ (%)	$\bar{I}_0$ (bit)
normal spectra	82	85	0.379
logarithmic spectra	79	84	0.335
binary spectra (threshold 0.05%)	73	78	0.216
binary spectra (threshold 5%)	76	75	0.210
normalization to total ion current	78	86	0.347
normalization to local ion current (window: 7 mass units)	83	85	0.390
significant spectra	77	82	0.291
significant spectra binary encoded	75	77	0.235
significant spectra normalized to local ion current (window: 7 mu)	75	80	0.257
2 largest peaks in intervals of 14 mass units	80	84	0.343
1 largest peak in intervals of 7 mass units	74	86	0.290

List of molecular structures corresponding to class 1 in each training:
double bond C=C,
double bond C-4=C-5,
hydroxysteroid,
oxosteroid,
estrane- or androstane-type,
3-hydroxysteroid,
3-oxosteroid,
oxygen function at C-11,
oxygen function at C-17,
OH in side-chain at C-17,
CO in side-chain at C-17,
20-oxopregnane,
20-hydroxypregnane,
carboxy group   (mu: mass unit)

the K-nearest neighbour method. Linear separators did not work well, especially those based on learning machines.

Recognition of the molecular weight by a multicategory classifier (trained with a least squared error method) was unsatisfactory [160].

A method for the recognition of the molecular ion in a mass spectrum was reported by Nägeli [217]. Classifiers have been calculated from "difference mass spectra" (mass differences to the molecular weight are used instead of ion mass numbers) applying the learning machine. A predictive ability of 86 % has been reported for a correct indication whether an assumed molecular weight is true or wrong.

13.1.4. Recognition of Molecular Structures

The most challenging task of pattern recognition applications in mass spectrometry is the automatic recognition of molecular substructures in a molecule. In a great number of papers classifiers for this purpose were developed by the learning machine method; typical predictive abilities are 70 to 95 %.

Jurs, Mc Gill, et. al.: [1, 132, 197],
Justice, Isenhour, et. al.: [133, 135],
Kent, et. al.: [141],
Mathews: [192, 193],
Nägeli: [217],
Varmuza, Rotter, et. al.: [163, 248, 310, 311, 315],
Volkmann: [318],
Wilkins et. al.: [279, 332].

Modifications of the learning machine, appropriate preprocessing, and feature selection improved the classification results. Use of cross terms (which take into account interactions between two mass numbers) accelerated the training but had less influence on the predictive abilities [124]. The introduction of a width parameter into the learning machine slightly improved the predictive ability and the absolute value of the scalar product could be used as a measure of confidence [320]. A Fourier transformation of the mass spectra had almost no effect on the predictive abilities [122, 321].

Combination of the mass spectrum, infrared spectrum, melting point, and boiling point to a single pattern vector improved the results for the recognition of a carbon-carbon double bond, but had almost no influence on the recognition of an ethyl or vinyl group [130].

Interpretation of mass spectra by distance measurement to centres of gravity has been reported by Justice and Isenhour [133, 135], Mathews

[193], Rotter and Varmuza [248, 250, 315], and Volkmann [318].

The method of linear regression was used to compute binary classifiers which can recognize chemical structures of steroid molecules [251, 314].

Wilkins et. al. have reported the application of simplex optimization to the training of classifiers for the recognition of 11 molecular structures [24, 165, 242, 243]. The best decision vectors found by the authors are completely tabulated [165].

Successful applications of the K-nearest neighbour method have been reported by
Burgard, et. al.: [33, 34],
Justice, Isenhour, et. al.: [133, 135],
Lowry, Ritter, Isenhour, et. al.: [172, 242, 243, 246],
Varmuza, Rotter: [248, 305].
The Euclidean distance between mass spectra in the pattern space was shown to be a good measure of chemical and structural similarity at least for low molecular weight organic compounds. A good clustering was found for alkanes, alkylbenzenes, aliphatic compounds containing a carbonyl group, compounds bearing a (substituted) amino group or a phenyl group and phenols [305].

Ziemer et. al. [362] utilized the KNN-method for the identification of amino acid sequences in polypeptides. A multicategory classification was applied to identify separately the N- and C-terminal amino acids in dipeptides. 40 structural classes were defined representing the 20 commonly occurring amino acids in both the N- and C-terminal position. The classification accuracy for 86 pentafluoropropionyl dipeptide methyl esters was 100 % in the N-position and 97 % in the C-position.

Parametric methods were investigated by Franzen and Hillig [86, 87, 108], and for a specialized problem by Vink et. al. [317].

Early optimism by Stonham, Aleksander, et. al. [282, 283, 284, 285] about the usefulness of adaptive digital learning networks could not be confirmed by Wilkins et. al. [280].

Piecewise multicategory classifiers have been used for the recognition of C=C double bonds because the data set was not linearly separable [89]. Several other pattern recognition methods have been used with more or less success to develop classifiers that recognize molecular structures [9, 12, 134, 142, 144, 173, 184, 197, 211].

Comparisons of different pattern recognition methods are problematic but generally indicate that the best methods are
        K-nearest neighbour classification,
        linear regression,
        simplex optimization.

Less satisfactory results have been obtained by distance measurement to
centres of gravity, learning machine or other methods. Some sequences
of methods with decreasing quality have been reported :

```
KNN > learning machine > centres of gravity                     [135]
learning machine > centres of gravity               [193, 315, 318]
learning machine = simplex > learning network                   [279]
simplex = KNN > learning machine > centres of gravity      [242, 243]
regression = KNN > learning machine                             [251]
```

Comparitative studies revealed that good classification results can
be expected for the following chemical classes of compounds (parameters):

benzenes, aldehydes or acetals, phenols, thiols or thiolethers,
amines, nitriles [165];

phenyl, three or more double bonds, carbon-to-hydrogen ratio,
presence of nitrogen [135];

presence of N, NH_2, O, OH, Cl, C=C, C≡C in a molecule [318].

Less satisfactory results were obtained for ethers, alcohols, acids,
esters, amides [165] and for number of carbon atoms, and for the
presence of a double bond [135].

Additional application of chemical knowledge to the selection of
features or to the classifier construction has improved the classification results [193]. A comparison between pattern recognition methods and
a sophisticated interpretative library search system for mass spectra
("STIRS" [39, 422]) has indicated some superiority of the STIRS-system
[172, 202, 332]. A decision tree pattern recognition was recommended by
Meisel et. al. [205] as a supplement to library search.

13.1.5. Chemical Interpretation of Mass Spectral Classifiers

Positive components of the weight vector should correspond to mass
numbers which are characteristic of the molecular structure to be classified (if it is defined that a positive scalar product indicates the presence of that molecular structure). Although it was not possible to in-
terpret in detail all weight vector components, the relationship was
principally confirmed by some authors.

Jurs [121] reported a weight vector with 31 components for the recognition of oxygen in a molecule. 14 components were positive and 10 of
them correlated with typical oxygen containing fragments. A similar
interpretation was given for a weight vector that recognizes the presence
of nitrogen in a molecule.

Another interpretation of weight vector components trained by a learning machine has been reported by Varmuza et. al. [311]. A table of key fragments [409] compiles a set of 24 mass numbers for typical oxygen--containing fragments; a weight vector trained for the recognition of oxygen in a molecule contained positive components at 21 of these mass numbers. A weight vector trained for the recognition of alkanes contained distinct positive components at those mass numbers (e.g. 51, 57, 85) which are characteristic of alkanes. The same effect occurred in weight vectors for the recognition of cycloalkanes (mass numbers 42, 56, 83, 97, 112), alkylbenzenes (mass number 91), and compounds bearing a oxogroup at a benzene ring (mass number 105).

Mahle and Ashley [184] have reported the top ten mass numbers with positive components for an oxygen classifier and an amine classifier.

Less satisfactory correlations between weight vector components and mass numbers were found by Nägeli [217].

Justice and Isenhour [136] used factor analysis to correlate mass numbers with seven different molecular structures.

Correlations between components of a classifier and characteristic fragments reveal that the learning machine is able to select automatically relevant mass numbers without any mass spectrometric knowledge. However, up to now it has not been reported that new mass spectrometric knowledge could be deduced from pattern classifiers.

13.1.6. Simulation of Mass Spectra

Jurs, Schechter, and Zander [125, 260, 261, 359] applied a set of binary classifiers to predict the mass spectra of compounds from the known molecular structures. This approach is completely empirical and does not assume any mass spectrometric knowledge of fragmentation of ions.

Each compound was described by a set of about 100 descriptors. The descriptors were generated from a fragmentation code or a coded structure (connection table). The set of descriptors is used as a pattern vector that characterizes a compound. Learning machine [261, 359] and an iterative least-squares method [260] have been used to train binary classifiers that predict mass spectral peak presence or absence at certain mass numbers. 60 classifiers for 60 mass numbers predict whether the peak is greater than 0.5 % of the total ion current. (The total ion current is the sum of all peak heights in a spectrum.) For 11 of these mass numbers

additional 22 classifiers were trained for two other intensity cutoffs
(0.1 % and 1 % total ion current). This set of 82 classifiers was used to
calculate a simplified mass spectrum from the known molecular structure.
For a data set of 600 low molecular weight compounds of chemical formulas
$C_{3-10}H_{2-22}O_{0-4}N_{0-2}$ a satisfying agreement with the actual mass spectra
was reported with predictive abilities between 71 and 97 %. Correlations
between the required descriptors and the fragments responsible for a particular mass spectral peak were discussed by the authors [359].

The method was proposed for an automatic interpretation of unknown
mass spectra. If a set of possible compounds is assumed for the unknown,
artificial mass spectra can be calculated and compared with the mass
spectrum of the unknown. The compound associated with the artificial
spectrum which compared most closely would then be assumed to be the compound producing the unknown spectrum. However, successful generation of
suitable classifiers for more mass numbers and more intensity cutoffs
would require very large data sets.

13.1.7. Miscellaneous

Concrete applications of pattern classifiers to the interpretation
of mass spectra has been reported by Wilkins et. al. [138, 165]; An on-
-line pattern recognition system was implemented at a minicomputer as a
diagnostic aid to the practicing spectroscopist. Varmuza [308] described
a computer program running on a time-sharing computer in connection with
an on-line minicomputer. Mc Gill [197] applied several pattern recognition methods from the "ARTHUR" program package [78] to mass spectra.

Numerical values of classifiers have been reported only in a few
papers [129, 163, 165, 192].

A promising field of pattern recognition applications involves preliminary classification of mass spectra at a GC/MS analysis or at a frac-
tionated evaporation of a sample [47, 49, 86, 95, 270, 271, 307]. It is
usually impossible to compare hundred or more spectra with a large spectral library. But a classifier may extract those spectra which stem from
a certain interesting class of compounds. For this purpose, the classifier has to be optimized for best recall rather than for best precision
in order not to lose interesting spectra [47]. In a subsequent step, the
selected spectra may be compared with a library to identify the compounds. The result of classification for a series of spectra was graphi-
cally represented in a similar way as a gas chromatogram [307]. The pro-

duct of the classification result (scalar product) and the concentration
of the compound (total ion current) was plotted against time (equivalent
to spectrum number). Positive peaks indicate compounds containing a certain molecular structure and negative peaks compounds without this structure. The graph corresponds to a gas chromatogram registered with a
(fictitious) detector sensitive only to a certain chemical structure.
By analogy with mass chromatograms, this graph was called "classification
chromatogram".

A real-time interpretation during the GC/MS run by several classifiers and recording of the classification chromatograms in real-time on a
multichannel recorder or display was proposed by Franzen [86].

Grouping of compounds in the classical way by the use of substructures or chemical reactivity may be inappropriate for pattern recognition
purposes [141]. Natural grouping of the compounds in the mass spectral
pattern space is probably different from that of the usual classes of
compounds. Heller et. al. [105, 106, 268] investigated the clustering of
mass spectra from alkylthioesters. Rotter [248] did the same for mass
spectra of steroids. No definitive useful facts could be concluded from
these cluster analysis up to now.

A series of papers - related to pattern recognition - about the
application of factor analysis to mass spectra has been reported by
Rozett and Petersen [225, 252 - 256].

13.2. Infrared Spectroscopy

Pattern recognition methods have been successfully applied in about
twenty papers to the interpretation of infrared spectra. In these works
classifiers for the recognition of chemical classes have been developed
for organic compounds.

Pattern vectors are directly computed from the infrared spectra by
digitizing at 0.1 micron intervals, and each interval corresponds to one
vector component. The number of intervals (dimensions) is about 130. The
absorption values have been used in three different ways to calculate
vector components:

1. The vector components are equivalent to the measured absorptions.
2. Numerical values of the vector components are restricted to 4 to 10
discrete steps.

3. For binary encoded infrared spectra each vector component is set to 1
if a peak occurs in the interval and is otherwise set to 0. Despite the
extreme data reduction, good classification results have been obtained
with this preprocessing.
The vector components are usually assumed to be independent.

The learning machine approach has been used by Kowalski et. al.
[159] to compute binary classifiers for the recognition of 19 chemical
classes (training set: 500 spectra, prediction set: 3500 spectra, overall predictive ability averaged over 19 structures: 73 - 87 %). The
learning machine method was found to be superior to a classification method based on average spectra. Unsatisfactory results have been obtained
if the population in the classes to be separated differed significantly.

Somewhat better results were reported by Lidell and Jurs [167, 168]
for a smaller set of only 212 infrared spectra. Several modifications
of the learning machine and appropriate feature selection gave overall
predictive abilities of 92 - 98 % (average of seven chemical classes).
The positive components of the weight vectors agreed quite well with
characteristic bands in the infrared spectra [167, 168, 233]. A drastic
reduction of the number of features was possible without a great loss
of predictive ability (Table 12).

TABLE 12. Interpretation of infrared spectra by binary classifiers based
on the learning machine. The overall predictive ability P decreases
only slightly if the number of features is reduced from 128 to 10 [233].

class	overall predictive ability (%)	
	128 features	10 features
carboxylic acids	96	93
esters	97	92
primary amines	95	93

Comerford et. al. [52, 53] reported that the learning machine did
not converge if the training set contained more than approx. 550 spectra.
Overall predictive abilities were 89 - 100 % for six different chemical
classes. The results are improved if infrared and Raman spectra are con-
catenated into one vector.

Extensive studies about the behaviour of the learning machine have been performed with simulated infrared spectra [233].

Recognition of carboxylic acids by the KNN-method and distance measurements to centres of gravity were investigated by Woodruff and Munk [354]. Predictive abilities for binary encoded infrared spectra were 64 - 70 % and could be improved to 70 - 88 % by consideration of shoulders in the spectra. However, a computer program based on chemical interpretation rules yielded significantly better results (91 % correct).

Penca, Zupan, and Hadzi [221] used a hierarchical tree for a discrimination between seven groups of carbonyl compounds and achieved 80 - 83 % correct results.

Several classification methods for binary encoded infrared spectra have been tested by Isenhour et. al.. The library of 2600 spectra contained 200 spectra from each of 13 mutually exclusive chemical classes. Molecular formulas were $C_{1-15}H_xO_yN_z$ and 139 intervals in the infrared region from 2.0 to 15.9 μm were used. Feature selection for these data was investigated by Lowry [170]. The most important wave lengths were found at 4.5, 5.8, 3.7, 5.9, 3.0, 3.6, 3.8, 3.1, 6.5, and 4.4 μm. A reduction to only 32 features lowered the predictive ability by only 2 %. Distance measurements gave predictive abilities of 90 % for the discrimination between two chemical classes, and 82 % for the recognition of one out of 13 classes [171]. Similar predictive abilities were found for a maximum likelihood classifier [176, 244]. A considerable improvement was achieved by the additional use of peak absences in the intervals [176] and by the additional application of correlation terms [355]. In Table 13 are compiled results for a multicategory KNN-classification; a consideration of 3 or 5 neighbours instead of only one yielded poorer results. Multicategory classification of all 13 chemical classes was also performed by distance measurements to centres of gravity; predictive abilities between 27 and 93 % (average: 59 %) must be compared with random guess (100/13 is approx. 8 %) [351, 352]. A comparison [356] of five multicategory classification methods for binary encoded infrared spectra revealed the following sequence of decreasing performance:

maximum likelihood > distances to centres of gravity > KNN-method with Tanimoto distances = KNN-method with Hamming distances > classification by mean vectors.

Drozdov-Tichomirov [76] used a very laborious method with potential functions to distinguish between compounds with and without a carbonyl group (training set: 194 compounds with 92 in each class, prediction set: 60 compounds, overall predictive ability: 95 %).

TABLE 13. K-nearest neighbour classification of 13 chemical classes
from binary encoded infrared spectra tested with the leave-one-out
method. The Tanimoto distance was used because it gave slightly better
results than the Hamming distance. P_1 and P_2 are the predictive abilities for class 1 and class 2, $\bar{P}$ is the average of P_1 and P_2; all values
are averaged over all 13 chemical classes. Notice the extremely poor recognition of class 1 if spectra of the other class occur much more frequently [353].

| number of spectra | | predictive abilities (%) | | |
class 1	class 2	P_1	P_2	$\bar{P}$
200	2400	55	96	76
200	200	86	66	76

Hierarchical clustering was applied to find out those wavelengths in
binary encoded infrared spectra with the smallest correlations. The result was utilized for a library search [104]. Methods similar to pattern
recognition have been used for the detection of atmospheric constituents
by a CO_2 laser [214, 215].

Short reviews about pattern recognition applications in infrared
spectroscopy have been given by Isenhour and Jurs [117], and Kowalski
[148].

13.3. Raman Spectroscopy

Only a few works have been performed on pattern recognition applications in Raman spectroscopy. This is probably due to the lack of suitable computer-readable spectra. Methods of coding and classification
would be the same as used for infrared spectra.

Schrader and Steigner [263] used averaged Raman (and infrared)
spectra for the classification of molecular structures of steroids.
A set of 42 structures was investigated using spectra from 70 steroids.
The spectra were divided into intervals of 10 cm^{-1} width and the intensities of adjacent intervals were also considered in the calculation of
the pattern components.

- 29, 333, 334, 336]. In the first studies, 500 proton-noise-decoupled
^{13}C-NMR spectra recorded with two different instruments and with eight
different solvents were used. The pattern vectors were generated by divi-
ding the spectra into 200 intervals with a width of 1 ppm. Different preprocessing methods (binary encoding, normalization to the sum of all peak
heights, Hadamard transformation, etc.) have been tested in connexion
with the learning machine; however, no obvious superior preprocessing
technique was found [29, 336]. The performances of the classifiers for
seven chemical classes differed markedly, e.g. the class "carbonyl" was
less recognized than a more precisely defined class "aldehydes and ketones". The averaged overall predictive ability for the classification of
three chemical classes (aliphatic alcohols, carbonyl compounds, phenyl
compounds) was similar for three different types of spectra: 84 % for
^{13}C-NMR spectra, 95 % for infrared spectra and 86 % for mass spectra.

Somewhat higher predictive abilities could be achieved by the application of a committee machine consisting of five classifiers [334]. These
classifiers were calculated by a learning machine using five different
preprocessing methods. A simple majority voting produced 86 % correct
answers (average for seven chemical classes and 62 compounds). The committee machine always yielded better results than any single classifier.

Classifiers which had been trained with simulated free-induction
decay data did not reveal satisfactory classification performance for
actual spectra [28].

The complete numerical data were given of three decision vectors for
the recognition of compounds bearing a phenyl, carbonyl and methyl group
[27]. These classifiers were calculated by a simplex optimization which
provided significantly better results than a learning machine. Classification using these decision vectors is carried out by computation of
scalar products. A positive value predicts the presence of the group in
question. Chemical shifts less than tetramethylsilane (0.00 ppm) are included in the first component of the decision vector; chemical shifts
greater than 199.00 ppm are included in the last (200 th) component. The
other features correspond to consecutive 1 ppm intervals from 0.00 to
198.99 ppm. ^{13}C-NMR spectra are binary encoded ("peak - no peak").
A maximum number of 102 features were selected by use of the Fisher ratios. The authors emphazised that an averaged predictive ability of 76 %
which was found with 2000 unknowns is a level which should be of substan-
tial help in spectral interpretations when combined with additional information.

Less satisfactory results have been reported [333] for binary classifiers which were computed by the simplex method. From a collection of

3782 binary encoded ^{13}C-NMR spectra a training set of 300 spectra has
been selected for the computation of classifiers for 24 chemical classes.
Simpler methods like a Bayes classifier or a maximum likelihood classifier yielded significantly better results than classifiers which were
optimized by a simplex algorithm.

Woodruff et. al. [357] compared several pattern recognition methods
with the recognition of nucleosides, carbohydrates and steroids from
^{13}C-NMR spectra. Table 14 shows the averaged recognition abilities for
2471 binary encoded spectra (the same spectra have been used in this work
for training and evaluation).

TABLE 14. Recognition of nucleosides, carbohydrates, and steroids from
^{13}C-NMR spectra by different pattern recognition methods. The recognition abilities for class 1 (denoted chemical class) and class 2 are
averaged values for all three chemical classes [357].

method	recognition ability (%)	
	class 1	class 2
KNN with Tanimoto distance	98	99
scalar product with mean vector	100	48
distance to centres of gravity	95	99
maximum likelihood	99	98

The SIMCA-method was used by Sjöstrom et. al. [266, 267] to distinguish between exo- and endo-2-substituted norboranes. Seven features
were generated from the ^{13}C-NMR spectra by using shift differences between the actual structure and the parent unsubstituted norborane. Exo
and endo compounds formed two well separated clusters. The results may
be used to classify other exo- or endo-2-substituted systems closely
related to norboranes. A similar method was applied to the detection of
ortho substitution in chlorinated diphenyl ethers [266].

13.5. Gamma-Ray Spectroscopy

The learning machine was successfully applied by Wangen and Isenhour [319, 322] to a semiquantitative determination of 17 light elements involving classification of the 14 MeV neutron induced gamma-ray spectrum. A training set of 160 pattern vectors (each with 64 components) was generated by mixing spectra of pure elements. A set of three binary classifiers for three concentration thresholds was computed for each of the elements B, N, Na, Mg, Al, P, Cl, K, Sc, Ti, V, Cr, Mn, Co, Ni, Cu, Zn. Figure 60 shows the application of these classifiers to the determination of the concentration range.

Average overall predictive abilities for the elements P, Cl, and B were 99 % for concentrations above 10 %,
95 % for concentrations between 1 and 10 %,
80 % for concentrations below 1 %.
Ni and Co gave the lowest predictive abilities.

Another application of pattern recognition methods to gamma-ray spectra was presented by Murrow [216].

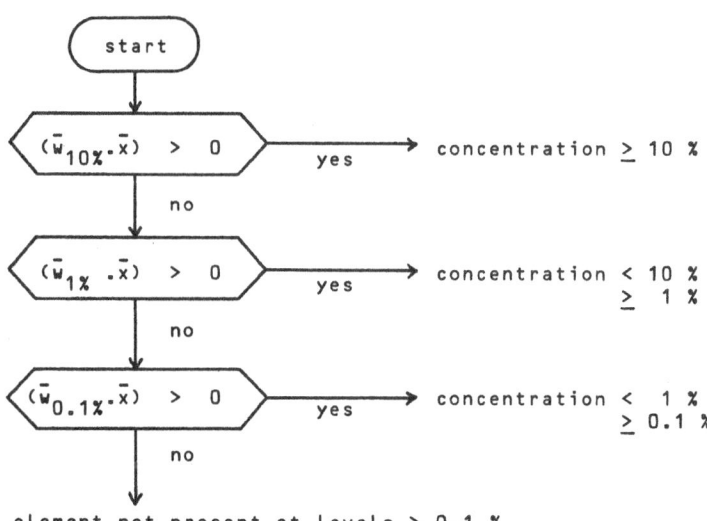

FIGURE 60. Semiquantitative determination of an element by classification of the gamma-ray spectrum (pattern $\bar{x}$). Three weight vectors $\bar{w}_{10\%}$, $\bar{w}_{1\%}$, and $\bar{w}_{0.1\%}$ are employed to determine the concentration range [322].

Comerford et. al. [53] combined infrared spectra and Raman spectra. Best results were achieved by concatenating the data from both sources and treating them as a single vector. Predictive abilities for esters, alcohols, ethers, compounds containing C=C double bonds and ketones ranged from 89 to 100 % (mean: 95 %). The decision vectors have been used for the assignment of vibrational frequencies.

13.4. Nuclear Magnetic Resonance Spectrometry

Pattern recognition techniques were used for the elucidation of molecular structure information from proton NMR spectra and from ^{13}C-NMR spectra. An excellent review was given by Wilkins and Jurs [335].

Kowalski and Reilly [161] started with a digitization of ^{1}H-NMR spectra in 2.5 Hz intervals (in the region 0 - 500 Hz relative to TMS) to generate pattern vectors with 200 components. The classification performance was poor because of the translational variance of the spectral patterns. A solution to the problem involves preprocessing the pattern vectors by an autocorrelation transform.

The autocorrelation spectrum A(t) of a digitized NMR spectrum B(f) is given by equation (150).

$$A(t) = \sum_f \{ B(f).B(f+t) \} \tag{150}$$

Simulated NMR spectra have been used to demonstrate that linear classifiers (computed by a regression analysis) can recognize ethyl, n-propyl, and isopropyl groups. In a subsequent paper [152] it was shown that a training set of 99 NMR spectra was not linearly separable. Consequently, an overall predictive ability of only 45 % was obtained by the application of a learning machine. The KNN-method yielded significantly better results (93 %) with the same data.

Autocorrelation was also utilized by Zupan et. al. [364] for the generation of pattern vectors from 552 computer simulated NMR spectra. The classification problem was the determination of the anomeric configuration of D-ribofuranosyl nucleotides. A learning machine operating with 100 spectra produced classifiers with a predictive ability of 76 - 100 %.

The application of several pattern recognition methods to the interpretation of ^{13}C-NMR spectra was investigated by Wilkins et. al. [26 -

13.6. Combined Spectral Data

Usually, no single spectral technique is sufficient to answer complex questions such as a chemical structure determination. Jurs et. al. [130] combined mass spectra, infrared spectra, melting point, and boiling point to pattern vectors. Combined patterns have been formed for 291 compounds with chemical formulas $C_{1-10}H_{1-24}O_{0-4}N_{0-3}$. Each pattern contained 132 components from mass positions and 130 components from absorption wavelengths. The relative contributions of the two types of data grossly affected the classification behaviour. Therefore, the data were normalized so that each data source contributed equally to the total amplitude of the pattern set.

The learning machine was used to train classifiers for the recognition of ethyl groups, vinyl groups and C=C double bonds. A significant improvement of the classifier performance with combined spectral data was only found for the determination of double bonds. The number of features could be reduced to 20 without a decrease of the predictive ability.

A combination of the infrared spectrum and the Raman spectrum to a single pattern vector has been successfully applied by Comerford et. al. [53] (Chapter 13.3).

An alternative approach to the combination of data from different sources would be successive applications of classifiers as described in Figure 51 (Chapter 11.4).

14. Chromatography

14.1. Gas Chromatography

The number of stationary phases available for the use in gas liquid chromatography (GC) is large. The chemist is confronted with the problem to make a choice among this large number. Pattern recognition methods can be used for an objective description of the similarity of GC phases. Extensive work in this field has been done by Isenhour, Massart, and Wold [189, 240, 340].

Mc Reynold [423] characterized 226 liquid phases by the retention indices for ten solutes relative to squalane. Each of the GC phases is therefore described by ten measurements (features) and can be represented by a point in a 10-dimensional space. A numerical analysis of these multivariate data may be carried out with different scopes [98, 189, 240].

1. To find out the number of solutes (perhaps less than ten) that are necessary to characterize the liquide phases. An appropriate pattern recognition technique is feature selection.

2. Selection of a set of standard stationary phases. A cluster analysis will show which GC phases lie close together (and therefore display similar separating behaviour). Only one of several similar phases is put into a standard list of phases.

3. Substitution of one GC phase by a similar one or selection of two or more GC phases with different separation properties. The KNN-technique or a display method or a dendogram will detect the most similar or the most dissimilar phases.

4. A factor analysis may show the intrinsic chemical properties of the GC phases that mainly influence the separation process.

The general assumption for all pattern recognition approaches to these problems is that the distance between two phases in the 10-dimensional space is an appropriate measure of the chemical similarity. This hypothesis seems to be confirmed by the success of pattern recognition methods applied in this field. Usually, the Euclidean distance D is used. The coordinates ΔI_{ji} of phase j are given by the differences of Kovats indices for solute i at phase j and at squalane.

$$D_{AB} = \left[\sum_i (\Delta I_{Ai} - \Delta I_{Bi})^2 \right]^{1/2} \tag{151}$$

The summation is taken over all solutes i that are used to characterize the GC phases.

The application of the KNN-classification method and an exhaustive search for the best subset of features showed that four solutes were sufficient to characterize 226 GC phases [133, 166, 170, 175]. These solutes are 1-butanol, 2-pentanone, 1-nitropropane, and pyridine.

A cluster analysis of a set of 226 GC phases was performed by using the minimal spanning tree method [240] and the SIMCA method [340]. Most of the GC phases could be grouped into 13 to 16 clusters, each cluster containing only phases showing similar properties with respect to the ten solutes. The remaining 28 to 41 phases were classified as outliers being dissimilar to the phases in the clusters. Massart et. al. analyzed the same data by the hierarchical clustering method and found 30 clusters [69, 188 - 191]. Similar results have been obtained by use of the information theory [83].

Factor analysis of Mc Reynolds' retention index data yielded three important factors influencing the separation process [348]. The first factor is closely related to the polarity of the liquid phase, the second factor depends almost solely on the solute and the third factor is related to interactions with hydroxy groups in the solute. Similar results were reported by other authors [112, 174, 196, 241].

A comparison of some classification methods used for stationary phases was reported by Fellous et. al. [84].

Retention index data can be employed to recognize molecular structures of unknown monofunctional compounds. This was principally demonstrated by Huber and Reich [113, 237]. Retention data of 199 compounds on ten different stationary phases were utilized for cluster analysis, KNN--classifications and the computation of classifiers with the learning machine. The minimum number of stationary phases was only two for the classification of aromatic compounds, four for alcohols and 13 for aldehydes and ketones. A two-step classification procedure was developed. The first step is the determination of a correction parameter $100\,n_{sk}$ for the retention index I. The skeleton number n_{sk} is defined by the number of atoms in each compound, excluding hydrogen atoms. This number was determined by a multicategory classifier developed by a learning machine. The modified data $I - 100\,n_{sk}$ are used in a second step for the

determination of functional groups. The modified data showed a significant better clustering of compounds than retention indices.

14.2. Thin Layer Chromatography

The solvents used in thin layer chromatography (TLC) can be classified by the same methods as used for stationary phases in gas chromatography. The separation characteristics of parallel TLC runs with different solvents should differ as much as possible. Massart et. al. [67, 68,
187] applied pattern recognition methods (numerical taxonomy and cluster
analysis) for an objective characterization of the solvents.

One application deals with the selection of optimum solutes for a
TLC separation of 26 synthetic food dyes. Hierarchical clustering was
employed to reduce the number of solutes from ten to four [187].

15. Electrochemistry

Applications of pattern recognition methods to polarography were extensively investigated by Perone et. al.. The first study [294] deals with the qualitative analysis of solutions of Cd^{2+}, In^{3+}, and Sb^{3+} ions by stationary electrode polarography. Polarograms were digitized at 2 mV intervals to generate pattern vectors of synthetic mixtures. The learning machine was used to compute classifiers that recognize the presence or absence of individual metal ions. A good separability of the classes was found but predictive abilities were only moderate. Main problems are the (normal) fluctuation of electrochemical data and the sometimes severe peak overlap. Successful classification was obtained only for a peak separation of at least 40 mV and for maximum concentration ratios of 10:1.

Another paper [296] deals with the recognition of seven metal ions (Cd^{2+}, Ni^{2+}, Co^{2+}, Pb^{2+}, Sb^{3+}, Tl^{+}, Cu^{2+}). However, classifiers computed from synthetic polarograms did not work satisfactorly on actual measured curves. A non linear mapping of the data showed that no compact clusters exist; especially the patterns for Co^{2+}, Cu^{2+}, and Ni^{2+} lie far outside of the clusters. Therefore, it is not surprising that the best classification result (79 % correct) was obtained by the KNN-method.

Pattern recognition methods were applied to the determination of the multiplicity of peaks in stationary electrode polarograms [295]. A set of 133 features describing the peak shape were derived from polarograms. After a feature selection (by using the learning machine) only 22 features were necessary to obtain a predictive ability of 90 % for a discrimination between singlet and doublet peaks. A selection of the three most important features by using the KNN-method for feature selection further improved the overall predictive ability to 98 % [228]. Recognition of the multiplicity may be useful to test the purity of a sample and is in principle applicable to wave forms obtained from any source. Extensive theoretical studies on the same topic were reported [297]. An on-line KNN-classification of peak multiplicity was reported [70]; real voltammograms of seven different metal ions were compared with a training set of real and theoretical data.

Pattern recognition methods have been proposed for the determination of the Cl^{-}/Br^{-} ratio by use of ion-selective electrodes [37]. Furthermore, multicategory classifiers have been computed for the quantitative determination of Cd^{2+}, Pb^{2+}, and Tl^{+} in molar concentrations between

10^{-8} and 10^{-6} by using anodic stripping voltammetry with a hanging mercury drop electrode [21].

Functional groups of organic compounds that are highly polar or un-
saturated are often reducible at a mercury electrode. Several pattern recognition methods were used to recognize four molecular structures (ali-
phatic-nitro, aromatic-nitro, aromatic aldehydes, and aliphatic-aromatic-
ketones) from voltammograms [32]. A set of 30 organic compounds were characterized by peak potentials, and by features derived from scan rate
dependence and from curve shapes. Somewhat surprising is the result that
a single feature (peak potential) was able to classify 93 % of the 30
compounds correctly.

Kinetic parameters of electrochemical reactions have been estimated
from voltammetric data by a KNN-classification [71].

The learning machine was used by Bos [20] for quantitative chemical
analysis by potentiometric titrations. Computer-calculated curves for the
titration of one strong base and two weak bases were utilized in the
training of multicategory classifiers. Prediction of concentrations (with
an error of $\pm$ 1 %) of the strong base was performed with patterns generated from the titration curve themselves; for the weak bases, it was advantageous to use the first derivative of the titration curves. The method was also successfully applied to actual samples.

16. Classification of Materials and Chemical Compounds

16.1. Technology

The origin of materials can often be classified by analyzing the contents of trace elements. The pattern of concentrations is caused by the raw materials which are used and by the production process, e.g. the trace elements in paper samples stem from wood, process water and other materials and substances used during the production. Duewer and Kowalski [79] used the concentrations of ten elements (Na, Al, Cl, Ca, Ti, Cr, Mn, Zn, Sb, Ta) - determined by neutron activation analysis - to characterize a paper sample. A set of 119 different sheets of papers from nine manu- facturers have been classified by several pattern recognition methods. Classifiers were computed to recognize both the manufacturer and the paper grade. The concentration of aluminum was found to be very charac- teristic of paper samples. A simple feature by feature plot (concentra- tion of aluminum versus concentration of manganese) revealed well defined groups of manufacturers. All attempts to find natural groups other than paper grade and manufacturers were unsuccessful.

Simon et. al. [265] applied electrothermal atomization atomic ab- sorption spectrometry to the determination of the concentrations of ten elements (Cu, Mn, Sb, Cd, Cr, Co, Ag, Pb, Mg, Fe) in 19 paper samples. Hierarchical clustering showed that only six features (Cu, Mn, Sb, Cr, Co, and the density) were needed to identify 16 of 19 papers correctly. The classification of paper samples by pattern recognition methods may be useful in forensic problems.

Another presumptive application of pattern recognition techniques is the interpretation of quality control data. Two such problems were described by Kowalski [147]. Quality control measurements were performed to characterize production items of an explosive in one case and of beryllium parts in the other. Pattern recognition methods may be useful for the evaluation of specification limits. If no classification method is capable of distinguishing between good and poor items, then it may be supposed that the data (quality control measurements) are insufficient. Feature selection and combination of features may indicate those measure- ments which provide most information on the separation into good and poor items [148].

Recognition of three electrical <u>fire-hazard classes</u> by using the
learning machine was examined by Weisel and Fasching [327]. A set of
13 features was derived for 47 chemical compounds from physical measurements and structural information. However, only 67 % of the compounds
could be classified correctly.

The mass spectra of a large group of pure compounds typical of those
found in <u>gasoline</u> have been used by Tunnicliff and Wadsworth [300] to
derive a set of weight vectors. These weight vectors were used to determine average properties (hydrocarbon type, average molecular weight)
of gasoline samples.

Kuo and Jurs [164] successfully applied pattern recognition methods
to the semiquantitative determination of the <u>chlorine dosage</u> necessary
for the treatment of water in a municipal water plant. A set of 17 water-
quality descriptors (e.g. turbidity, temperature, pH, alkalinity, hard-
ness, iron, manganese, ... total bacteria) was used to describe a water
sample. Each sample corresponds to a point in a 17-dimensional space.
The range of chlorine dosages was divided into seven intervals. For each
interval limit a binary classifier was computed using the learning machine and a training set of 30 samples. Each classifier splits the data
set into two classes, namely samples which exceed the limit of chlorine
dosage and samples which do not. A total of 104 samples for a two-year
interval was available. A test of the classifiers was performed with 50
randomly selected water samples. All classifiers are used to get a sequence of responses. The recommended chlorine dosage is given by the interval limit where the response changes from negative to positive (Chapter 2.6.3, Table 4). 84 % were classified into the correct interval of
chlorine dosage; in 94 % the chlorine dosage was within ± 1 interval of
the correct dosage. This study may be the first step to an automatic
chlorination system for water. A related study deals with the classification of water quality [304].

Pattern recognition methods were successful in uranium prospecting
[23]. <u>Geological data</u> have been used to train a classifier that recognizes uranium. Correct recognition of 74 - 83 % has been reported for an
area not used in the training. Christie [41] has published some criti-
cisms about pattern recognition applications in geochemistry.

An interesting application of pattern recognition is the classification of agricultural areas on pictures taken from a satellite in different spectral regions.

16.2. Archaeology

Pattern recognition is a powerful tool in the identification of archaeological artefacts. A typical example of a pattern recognition application in chemistry is the classification of <u>obsidian</u> artefacts by Kowalski et. al. [162]. A total of 45 obsidian samples from different sources in northern California and 27 archaeological obsidian artefacts of unknown origin were analyzed by X-ray fluorescence spectroscopy. For each sample ten trace elements (Fe, Ti, Ba, Ca, K, Mn, Rb, Sr, Y, Zr) were determined in a concentration range of 40 - 1000 ppm. Each obsidian sample is therefore represented by a point in a 10-dimensional space. A cluster analysis by non-linear mapping and by an eigenvector projection [155] indicated four distinct clusters. The origin of almost all of the 27 archaeological samples could be classified by a visual examination of the two-dimensional map. Similar good results were obtained with the learning machine and with the KNN-classification. A human interpretation of the multivariate raw data without a computer gave the same results as the pattern recognition approach, but required several hours of tedious work. The same data were used for a detailed comparison of several pattern recognition methods [267]. It is interesting and instructive that a careful feature selection showed that only two features are necessary to classify all artefacts correctly by a simple feature by feature plot [228].

A similar technique of data analysis but without sophisticated pattern recognition methods was applied by Bird et. al. [19]. The concentrations of three elements (F, Na, Al) were determined by proton induced gamma-ray emission in 700 obsidian artefacts from 20 sources. Two-dimensional plots "Al versus Na" and "F versus Na" showed distinct clusters.

<u>Flint</u> samples from four different mines and from prehistoric workshops were analyzed by neutron activation analysis to determine the contents of 14 trace elements. A successful separation of the four origins was possible by a Bayes classifier, by classification with centres of gravity, and by factor analysis [66].

<u>Potsherds</u> were classified by hierarchical clustering using the concentrations of 14 trace elements determined by neutron activation analysis [220].

<u>Soapstone</u> fragments found in an old trade centre of the Vikings in northern Germany were analyzed by spark source mass spectrometry. A cluster analysis of the data showed five distinct populations. A comparison

with samples of Norwegian soapstone quarries demonstrated that one of the
populations originated from a newly discovered Viking soapstone quarry
in South Norway. Two other quarries that were supposed by some archaeolo-
gists were definitely ruled out [238].

Another impressive application of pattern recognition techniques
has been reported by Mc Gill and Kowalski [197, 199]. The origin of the
quarzite used for the Colossi of Memnon in Egypt was determined by using
a combination of pattern recognition methods (appropriate preprocessing
and feature selection, KNN-classification, learning machine, cluster analysis). The concentrations of 18 elements were determined by neutron activation analysis for 106 samples from seven sources of Egyptian quarzite
and for 56 quarzite samples of the Colossi. Feature selection showed that
three elements (Eu, Ca, Sc) were essential for the characterization of
the quarzite samples.

16.3. Food

Gas chromatograms can be used as a fingerprint of complex mixtures.
It may be supposed that the GC-profiles of foods are correlated with the
quality, origin, flavour, etc.

A classification of milk samples according to their origin (cows,
sheeps, goats) was reported by Smeyers-Verbeke et. al. [56, 269]. The
concentrations of 15 fatty acids in 20 milk samples were determined by GC
and used as pattern components. Classification performance of 85 - 100 %
was better than by a visual comparison of the chromatograms.

Pattern recognition techniques have been used for the recognition
of counterfeit whisky [259]. Chromatograms of whisky samples contained
17 distinct peaks. These peaks were used without further interpretation
to generate patterns. Only two features (probably GC-peaks from isoamyl-
alcohol and acetaldehyde) were needed for a complete separation of original whisky samples and counterfeit whisky samples by pattern recognition
methods.

A set of 45 pure peppermint oils from ten origins was correctly
classified on the basis of 12 features obtained from GC. The same method
was also successfully applied to binary mixtures of peppermint oils
[103]. Similar studies deal with the classification of coffee according
to flavour classes [231] and classification of the age of potato chips
[231].

16.4. Biology

Only a few remarks can be made here on the classification of biological materials. Chemical taxonomy [293] tries to characterize biological individuals (e.g. plants) on the basis of the amounts of various
chemical compounds found in several parts of the individual. Pattern recognition might be a useful technique to handle these data [343].

Bacteria can be characterized by diameters of clear zones around
various antibiotic disks. Mc Donald [181] used 12 different antibiotics
to classify eight types of bacteria and obtained an average of 85 %
correct classifications with the KNN-method.

Another approach to the classification of bacteria was investigated
by Meuzelaar et. al. [212]. Pyrolysis products of bacteria have been analyzed by mass spectrometry (electron energy 15 eV). Peaks at 40 most
characteristic mass numbers have been taken as pattern features. A non-
linear mapping of data obtained with "listeria bacteria" revealed two
distinct clusters that perfectly correspond to the two serotypes of these
bacteria. Similar results have been reported by Wold [347].

Ulfvarson and Wold [301] quantitatively analyzed 17 trace elements
in blood samples from 81 welders (working with stainless steel or with
aluminum) and from 68 non-welders. Using several pattern recognition methods, no statistical difference of the two groups was found. The same
data were employed in a detailed comparison of pattern recognition methods [267].

In tissue samples from 17 anatomic regions in pig hearts, the concentrations of 13 trace elements have been determined by emission spectroscopy. By using distance measurements to 17 centres of gravity, 65 %
of the specimens were classified correctly with respect to their anatomic
region [325]. Concentrations of trace elements in the human brain of normal persons and of persons with various diseases have been utilized for a
multivariate analysis. Sick persons could be correctly classified, however, the data set used was very small [64].

16.5. Chemistry

An illustrative example to demonstrate the applicability of pattern
recognition methods in chemistry was given by Kowalski and Bender [153].

Six properties of 68 <u>chemical elements</u> were utilized to determine whether the oxide of the respective element (higher valence) is predominantly acidic, amphoteric, or basic. Each element corresponds to a pattern point in a 6-dimensional space. The coordinates were characterized by the following properties: most important valence, melting point, covalent radius, ionic radius, electronegativity, and enthalpy of fusion. Autoscaling was applied to the measurements so that each property exerts the same effect on classification. None of the properties could separate the elements correctly when used alone. On the contrary, a non-linear mapping from the 6-space to a 2-dimensional representation showed a distinct separation of the acidic and basic groups and an overlap of the amphoteric group with the two other groups. A KNN-classificatiion (with three neighbours) assigned all acids and bases correctly [148]. Furthermore, the continuous property "acidity" is recognizable from the pattern recognition results [155]. Similar results for the same data have been reported by Lin and Chen [169]. A pattern recognition study on the periodicity of the chemical elements has been reported by Pratt et. al. [232].

Wold et. al. [267, 341, 343, 350] used the SIMCA method to distinguish between the <u>cis</u>- and <u>trans</u>-configurations of <u>unsaturated carbonyl compounds</u>. The compounds were characterized by seven features derived from ultraviolet and infrared spectra. Pattern recognition confirmed the results found previously by Mecke and Noack [204].

<u>Hydrides</u> with the chemical formula $ABH_n D_{4-n}$ (A: alkaline metal, B: metal of group IIIb, n: number of hydride atoms, D: substituent other than hydrogen) are either stable or unstable compounds. Strouf and Wold [286] used 28 variables derived from structural data for the classification of the stability. An averaged predictive ability of 80 % was obtained with the SIMCA method.

<u>N,N</u>-Dialkyldithiocarbamate <u>complexes</u> have been separated by several pattern recognition methods into five chemical classes with features generated from ^{13}C-NMR and infrared spectra [230].

X-ray valence band spectra (obtained by an electron microprobe) were successfully used to characterize <u>copper minerals</u> [74]. Nine features reflecting the peak shape were derived from the spectra. Non-linear mapping showed well separated groups in the 2-dimensional representation.

Further applications of pattern recognition involve the determination of simple crystal structures from thermodynamic data [93], prediction of chemical reactions [77, 114], and prediction of helical regions in proteins [65].

17. Relationships between Chemical Structure and Biological Activity

17.1. Pharmacological Activity

17.1.1. General Remarks

The world-wide research budget expended on the development of new and better drugs was estimated in 1978 to be approximately 2.10^9 dollars, and the cost of developing a new drug approximately 4.10^7 dollars [99]. A variety of methods for establishing relationships between chemical structures and their biological activities have been applied [97, 99, 185, 236]. Because no satisfying theoretical model exists for structure-activity-relationships (SAR), it is not surprising that a complete empirical approach like pattern recognition has found some interest. Because of the high costs of synthesis and tests of a compound, even a pattern classifier with a rather high error rate (e.g. 30 %) would be economical [107].

The most difficult problem in pattern recognition applications to SAR-classifications is the formulation of meaningful descriptors that describe the molecular structure and are correlated with the classification problem. The widely used concept of a linear, binary classifier assumes a linear relationship between the structural properties (pattern components) x_i and the biological activity.

$$\text{biological activity} \quad = \quad w_0 \quad + \quad \sum_{i=1}^{d} w_i x_i \qquad (152)$$

This is undoubtedly an oversimplification.

Expectations concerning the possibility of using results from pattern recognition to generate new chemical structures of a specific activity (drug design) have not been fulfilled up to now. Nevertheless, the reported results are generally promising and classification of drugs will perhaps be the most valuable application of pattern recognition methods in chemistry. It cannot be the scope of pattern recognition methods to replace biological tests but classification results may be useful to set priorities.

Excellent reviews of the application of pattern recognition techniques to structure-activity-relationships studies have been written by

- Kirschner, Kowalski : 1978 [145],
- Martin : 1978 [185],
- Stuper, Brugger, Jurs : 1977, 1979 [287, 288].

Therefore, only a brief survey of this theme is given in the next Chapters.

A modular program system especially suited for SAR-classifications
was developed by Jurs et. al. [127, 287, 288, 291]. The same authors
intensively studied the computer-assisted manipulation and usefulness
of various types of molecular descriptors [411].

17.1.2. Sedatives and Tranquilizers

Several papers deal with the binary problem of separating sedatives
and tranquilizers. While in the first papers a relationship between mass
spectrum and biological activity was supposed (Chapter 17.3), the basis
in recent papers are structure-activity-relationships.

Chu [42] used a set of 46 substructural fragments (called "augmented
atoms") to describe 30 sedatives and 36 tranquilizers. Classification
success rates of various pattern recognition methods were between 85 and
94 %.

A larger data set (79 sedatives, 140 tranquilizers) was applied by
Jurs et. al. [288, 289] to the calculation of classifiers by a learning
machine. A sophisticated preprocessing and feature selection of the raw
data improved the predictive ability to 90 %.

A set of 160 barbiturates was coded using 46 molecular descriptors,
including fragments, substructures, environmental descriptors, and a
molecular connectivity index [287, 288, 292]. Classifiers were developed
with a learning machine that recognizes the duration of the depressant
effect. Predictive abilities of 94 % were reported for correct classifications of the duration time "less than t minutes".

Molecular structures have been represented by "molecular transforms"
which are computed from the three-dimensional atomic coordinates of the
molecule using X-ray diffraction data [278]. Predictive abilities of 90 %
were reported for classifiers developed with the learning machine (data
set: 72 sedatives, 114 tranquilizers) [276 - 278].

17.1.3. Cancer and Tumors

Kowalski and Bender [156] investigated automatic classifications of 200 anticancer drugs previously tested for activity (87 of them being actually active). Each drug was represented by 20 structural features. Most effective features were:
- number of sulfur atoms / number of atoms,
- number of C-S bonds / number of carbon atoms,
- number of S-H bonds,
- number of C=C bonds / number of carbon atoms,
- number of carbon atoms / number of atoms.

Predictive abilities of approx. 90 % were reported for three pattern re-cogition methods (learning machine, least-squares-method, KNN-method). However, the selection of the set of drugs and the set of features was criticized later [194, 302].

Recognition of antitumor activity of 138 drugs was reported by Chu et. al. [43]. The original number of 421 features for each compound was reduced to 51 by automatic feature selection and by an heuristic weighting based on pharmacological experience. Classifiers were computed from 138 compounds using a learning machine and the KNN-method. Success rates between 83 and 92 % have been obtained with a set of 24 unknowns.

A statistical-heuristic method for screening large files of compounds for antitumor drugs was described by Hodes et. al. [109]. Molecular structure features were used as predictors of biological activity.

Carcinogenicity of 46 derivatives of p-dimethylaminoazobenzene and 39 derivatives of benz[a]anthracene was classified by Dierdorf and Kowalski [72]. Structural descriptors were derived from three-dimensional models; success rates were between 80 and 92 %.

Cancerogenicity of polycyclic aromatic hydrocarbons has been classi-fied by Norden et. al. [219] using the SIMCA method. A set of 32 compounds was represented by eight measured variables (absorption bands, ionization potentials), and 15 theoretical, non-measured variables (computed from molecular parameters like binding energies, electron-donor and electron-acceptor properties). Results showed two factors influencing the carcinogenicity but no obvious chemical interpretation of these factors was possible. A similar work by Dunn and Wold [81] deals with structure-carcinogenicity relationships for a series of 4-nitro- and 4-hydroxamino-quinoline 1-oxides. The class of active compounds could be well described by a model of the cluster; on the contrary, the inactive compounds were found to be spread over a large region of the data space.

Jurs et. al. studied structure-carcinogenicity relations of
N-nitroso compounds [40] and on a heterogeneous set of organic compounds
consisting of 12 structural classes [127]. An appropriate set of 15 to
28 calculated molecular descriptors was identified that support linear
classifiers able to separate carcinogenic and non-carcinogenic compounds.
Predictive abilities between 78 and 92 % demonstrate that pattern recognition methods can be useful for the prediction of carcinogenic activity.

17.1.4. Miscellaneous

A perceptron classifier was used by Hiller et. al. [92, 107] for the
recognition of the anticonvulsion activity of 1,3-dioxanes. The molecular
structures have been coded by substructural entities; predictive abilities of 68 - 76 % have been reported.

Adamson and Bush have described the application of pattern recognition methods to SAR-studies of penicillins [4], and local anaesthetics
[5,6].

Cluster analysis was used by Hansch et. al. [100] as an aid in the
selection of substituents in drug design.

Coding of compounds by substructures and relationships to biological activities were investigated by Cramer et. al. [58].

A set of 13 pressor agents was separated by an eigenvector plot
into two groups of weak and strong activity [35]. The compounds were coded by ten substructural features. A three-dimensional eigenvector plot
of 43 pharmacological active compounds showed distinct clusters for anti-
depressants and antipsychotics. A similar study by Menon and Cammarata
[209] deals with a set of 39 drugs (alpha- and beta-adrenergic agents,
cholinergic agents, and central nervous system stimulants). The major
pharmacological classes present were correctly identified and the phar-
macologically unrelated compounds eliminated.

Relationships between physical properties and the inhibition of
monoamine oxidase by aminotetralines and aminoindanes have been studied
by Martin et. al. [186].

A procedure for identifying clusters of similar individuals has been
developed by White and Lewinson [329] and applied to the investigation of
pharmaceutical data.

Gund [97] has reported a method of searching three-dimensional
molecular structures causing biological activity. The interaction between a receptor and a drug molecule is interpreted as a geometrical

pattern recognition response of the receptor to the drug.

An SAR-study of 37 beta-adrenergic agents using the SIMCA method has been reported by Dunn et. al. [82]. The pattern recognition method classified correctly 100 % of the agonists and 88 % of the antagonists. A semiquantitative determination of the activity was possible by using the parameters of the cluster model.

17.2. Odour Classification

An early application of a pattern recognition method to the prediction of olfactory qualities from physicochemical data was reported by Schiffman [262]. A set of 39 odorants was represented by 25 parameters (including Raman spectral information). Non-linear mapping revealed correlations of the chemical structure and smell.

A detailed study on the extraction of important molecular features of musk compounds using pattern recognition techniques was performed by Jurs et. al. [25, 288]. A data set consisting of 60 musk odorants and 240 nonmusk compounds were coded with 68 computer-generated structural descriptors (including fragment, substructure and geometric descriptors). A careful interpretation of the results (obtained by autoscaling, feature selection and learning machine using the programs of the ADAPT-system [291]) yielded a subset of 13 descriptors which are most effective for the classification of musk odorants. The results indicate that several molecular parameters rather than a single parameter are necessary to predict odour quality. Average predictive ability was 95 - 97 %; nine previously unused musk odorants were correctly classified.

A similar study deals with the classification of trigeminally active compounds [288]. A set of 12 molecular descriptors has been selected that are most useful for a semiquantitative description of the activities of 47 compounds.

Pattern recognition techniques have been applied by Mc Gill and Kowalski to determine the intrinsic dimensionality of smell [197, 198]. A set of 43 features was derived from physical and chemical data (IR-, UV-, NMR-spectra, molecular weight, melting point, boiling point, density, specific rotation, solubility in water and alcohol) for each of a total of 47 compounds. Feature selection and eigenvector plots indicate that there are only two meaningful axes in the 43-dimensional data space. The axes were found to relate to the molecule's electron donor ability and its directed dipole.

17.3. Spectra - Activity Relationships

The spectrum of a compound represents the molecular structure in the form of a complex code. Therefore, a relationship may be expected between spectrum and biological activity. Such a spectra-activity relationship assumes that the mechanism of biological activity is similar to the physical and chemical processes which produce spectra. This similarity is at least doubtful and the direct way of structure-activity relationships seems to be more promising. An attempt to correlate mass spectra with biological activity has led to violent replies in the literature. Controversies on this theme made many chemists very suspicious against applications of pattern recognition.

Ting et. al. [299] reported an attempt to correlate the mass spectra of 30 sedatives and 36 tranquilizers with their biological activity. The mass spectral peak intensities at 30 mass numbers were used for the characterization of the compounds. 83 % of the compounds were classified correctly by the nearest neighbour; and mapping showed well separated clusters of sedatives and tranquilizers.

Criticism at this work was rather destructive. E.g. for the same compounds a classifier has been computed by a learning machine which successfully recognized whether the name of the substance had an odd or even number of characters [51]. Another paper [223] questioned the independence of the set of compounds used because more than half of the tranquilizers were phenothiazines. Therefore, similar names of the compounds should indicate similar activity. Each compound was represented by a curious "character vector" (consisting of 26 components, the first component being the number of 'A' in the name and the second component the number of 'B', etc.). A success rate of 84 % correct classifications by the nearest neighbour was obtained with these "character vectors" ! Ting [298] replied that the name of a compound reflects structural features and therefore good classification results based on "character vectors" are reasonable. Furthermore, it was stated that "spectral results may offer a different view of reactivity than obtained from a consideration of the structure alone".

These examples show that improper use of pattern recognition methods makes it possible to obtain good classification results even for senseless questions. Normally, such effects occur if the number of spectra is a) too small, b) too different in the classes or c) does not exceed significantly the number of dimensions [251].

Abe et. al. [2] reported another study on the verification of correlations between mass spectra and biological activity. Several pattern recognition methods have been applied to a set of 17 analgesics and 16 antispasmodics. Predictive abilities of more than 90 % have been obtained by the KNN-method and by the learning machine. A set of 30 features and the "leave-one-out-procedure" was employed.

Brent et. al. [22] used mass spectral data to predict biological activities of analogs of trimethoprim by the KNN-method.

18. Clinical Chemistry

Some diseases can be diagnosed by a set of clinical laboratory tests rather than by a single test. Pattern recognition may be a suitable tool for an automatic interpretation of such data sets [56, 102, 183, 273]. Only a few remarks on this theme can be given here.

Koskinen and Kowalski [146] have reported a reevaluation of a clinical chemistry data set consisting of 22 laboratory tests made on 70 individuals. Eight of the individuals were healthy and the rest had a specific disease of the liver. A projection of the 22-dimensional data space into three dimensions showed an interesting cluster similar to a "three-legged starfish". The centre of the cluster contained only the healthy individuals while each of the arms contained only one of the liver diseases.

Other applications of pattern recognition methods to a computer-assisted diagnosis of liver diseases have been reported by Baron [14], Burbank [31], Ramsoe et. al. [234, 339], and Solberg et. al. [274, 275].

Blood cell data were analyzed with moderate success by using a learning machine, the KNN-method, and by cluster analysis to recognize deficiencies of glucose-6-phosphate dehydrogenase [182].

Classification of three groups of thyroid functional states has been described by Coomans et. al. [54—56]. Feature selection showed that only two laboratory tests instead of five tests are necessary for this classification, as previously supposed.

Different thyroid function tests were also investigated by Barnett et. al. [13].

Further pattern recognition applications deal with serum chemistry [210, 328], treatment of arthritis [331], diagnosis of cancer [30], shock [338], and myocardial infarction [38].

19. Environmental Chemistry

19.1. Petroleum Pollutants

Petroleum oils are released into aquatic environment by shipping, off-shore drilling operations, and catastrophic events. The problem of determining the source of an oil spill is difficult for at least three reasons: weathering, contamination, and analytical error.

A differentiation between biomass and petroleum material was reported by Lysyj and Newton [177]. Pyrolysis products of dried algae and of outboard motor oil showed different gas chromatograms. A more detailed investigation was described by Clark and Jurs [44]. Each gas chromatogram was represented by a 19-dimensional vector. All 19 vector components were derived from the original chromatogram:

- peak heights above the unresolved background of 12 normal alkanes (n-C_{14} through n-C_{25}), of pristane, and phytane;
- peak height ratios C_{17}/pristane, C_{18}/phytane, C_{17}/C_{18}, pristane/phytane and C_{17}/background.

A set of 42 gas chromatograms of petroleum samples was analyzed by several pattern recognition techniques. Predictive abilities of 87 to 100 % have been obtained by a learning machine and by KNN-classifications (with 1 - 3 neighbours). The KNN-method was more successful than adaptive, binary classifiers. The effect of various preprocessing methods and feature selection were discussed in detail. In a later study [45], the same authors used 80 gas chromatograms taken from a set of four crude oils before and after weathering. The chromatograms were coded with 13 descriptors each:

- peak areas for the normal alkanes for C_{16} through C_{25} plus pristane and phytane;
- and one descriptor characterizing the area of the unresolved background.

For some of the oil types a trained classifier using only chromatograms of unweathered oils could classify chromatograms of the same oil after weathering with 100 % accuracy. Bayesian classifiers and the learning machine were used in these classifications.

Infrared spectroscopy is the most frequently used method for the characterization of oil spills. Kawahara et. al. [139, 140] employed seven wave lengths for the calculation of 42 ratios of infrared absorbances. Eigenvector plots of 194 oil samples separated three groups of oil types. A classification success of 99 % was reported.

Similar results have been obtained by Mattson et. al. [195] with a set of 194 oil samples from six different types. Each infrared spectrum was described by 21 peak heights.

Baer and Brown [10] compared the infrared spectra from naturally weathered petroleum and artificial weathered petroleum. From a total of 21 samples, the sources of 18 oil spills were correctly identified by KNN-classifications.

The natural elemental composition as analyzed by <u>neutron activation</u> analysis was utilized by Duewer et. al. [80] for the characterization of oil spills. A set of 40 oils was artificially weathered using nine different methods. Neutron activation analysis was performed for all nine weathered samples and the original oil sample of each oil. A total of 22 elements was detected; seven elements (Na, Al, S, Cl, V, Mn, and Br) were detected in all samples. Several pattern recognition methods have been applied to the identification of the sources of the weathered oils [267]. Feature selection indicated that vanadium was the most significant element for oil source identification, followed by nickel and sulfur. These results are in agreement with chemical experiences in oil industry. A recognition rate of 96 % was reported for KNN-classifications using the 22-dimensional data set. Results obtained with the SIMCA method were superior to those obtained by all other methods. It was shown that weathering considerably affects the trace metal character of oils and therefore the classification method must be appropriate to "fuzzy" data sets. A lower bound estimate indicates that at worst 80 % of the oils considered can be correctly classified after having been subjected to the various artificial weatherings. The upper bond estimate ranges up to 100 % correct classifications.

Identification of oil spills by pattern recognition methods has also been performed by use of ultraviolet spectroscopic data [62, 63].

19.2. Atmospheric Particulates

The elemental composition of atmospheric particulates can be used for the identification of the origin of the particulates. A set of characteristic correlating elements may indicate, for example, crustal

origin or anthropogenic origin (industrial activities, automobile emission, power plants). Appropriate pattern recognition methods for detecting
characteristic sets of elements are cluster analysis, mapping and factor
analysis. Correlation coefficients r_{AB} for elements A and B were used
for the calculation of "distances" $D_{AB} = 1 - r_{AB}$ between the elements
[90, 222]. Hierarchical clustering and non-linear mapping group together
elements with small "distances" D_{AB}.

Detailed results of measurements in the city of Tucson (Arizona) and
outside the city have been described by Gaarenstroom et. al. [90];
results from the Boston metropolitan area have been reported by Hopke
et. al. [111] and those from Cleveland by Neustadter et. al. [218].

19.3. Miscellaneous

Classification of lake sediments by means of factor analysis and
hierarchical clustering has been reported by Hopke et. al. [110]. For a
set of 79 sediment samples, 32 characteristic properties were determined
(e.g. concentration of 15 elements, percent organic matter, criteria of
the particle size distribution, water depth). Cluster analysis detected a
single cluster for samples from the centre of the lake and three different clusters for samples from near the shore.

Peterson [226, 227] used pattern recognition methods for predicting
sulfur dioxide concentrations from meteorological parameters. For days
with consistent wind direction, 75 % of the calculated values agreed with
the values observed at several locations in the metropolitan St. Louis
(Missouri).

Several pattern recognition programs have been applied by
Wegscheider et. al. [326] to the interpretation of a set of trace element
concentrations in groundwater. The data set consisted of 149 samples with
up to ten measurements made on each sample. Four different origins of
these samples have been successfully visualized by display methods.

20. Classification of Analytical Methods

An analytical method can be represented by a point or by a region in a multidimensional "space of procedures". The coordinates correspond to the parameters of the method, like accuracy, time, cost, etc.. Kaiser [421] applied the information theory to the estimation of the "informing power" of analytical procedures. Pattern recognition methods have been proposed by Wold et. al. [36, 341, 343] for an objective evaluation of analytical methods. A data matrix is obtained by the application of methods to a number of real samples. Mathematical models were constructed for the clusters describing the methods under consideration.

Vandeginste [303] used binary classifiers for a retrospective decision between two alternative analytical methods, UV-Vis spectrometry and atomic absorption spectrometry. 237 problems were selected from analytical papers; 131 of the problems were solved by use of atomic absorption and 106 by UV spectrometry. The parameters describing the analytical problem can be divided into three groups: parameters related to a) the analytical results, b) the sample and c) the sample matrix. Classifiers were trained on a training set containing 157 problems and tested afterwards on the remaining problems. A predictive accuracy of 85 % was reported. The results proved that it is principally possible to select the proper analytical method from two alternatives by pattern recognition; however, further research will be necessary to determine which parameters best characterize an analytical method.

The application of pattern recognition methods to achieve an optimum selection of stationary phases in chromatography has been discussed in Chapter 14.

A classification based on taxonomy of separation methods has been reported by Giddings [91].

Part   C

Appendix

21. Literature

21.1. Pattern Recognition in Chemistry

1. ABE H.; JURS P.C.: ANAL. CHEM., 47, 1829 (1975)
2. ABE H.; KUMAZAWA S.; TAJI T.; SASAKI S.I.: BIOMED. MASS SPECTROM., 3, 151 (1976)
3. ADAMS R.A.; SEDGWICK R.D.: ORG. MASS SPECTROM., 9, 884 (1974)
4. ADAMSON G.W.; BUSH J.A.: NATURE, 248, 406 (1974)
5. ADAMSON G.W.; BUSH J.A.: J. CHEM. INF. AND COMPUT. SCI., 15, 55 (1975)
6. ADAMSON G.W.; BUSH J.A.: J. CHEM. SOC. PERKIN TRANS. I., 1976, 168 (1976)
7. ALBANO C.; DUNN W.J.; EDLUND U.; JOHANSSON E.; NORDEN B.; SJÖSTROM M.; WOLD S.: ANAL. CHIM. ACTA, 103, 429 (1978)
8. ANDERSON D.N.; ISENHOUR T.L.: PATTERN RECOGNITION, 5, 249 (1973)
9. ARTEMOVA F.SH.; ANDREEVA L.N.; POLYAKOVA A.A.; CHELNOKOV YU.V.; UNGER F.G.: ZH. ANALIT. KHIM. (ENGLISH TRANSLATION), 29, 2285 (1974)
10. BAER C.D.; BROWN C.W.: APPL. SPECTROSC., 31, 524 (1977)
11. BAILEY A.: IN PAYNE J.P.; BUSHMAN J.A.; HILL D.W. (EDS.): THE MEDICAL AND BIOLOGICAL APPLICATION OF MASS SPECTROMETRY, P. 93, ACADEMIC PRESS, LONDON (1979)
12. BAKER A.G.; CAMP M.; HUNTINGTON E.; PIKE W.T.; SHAW M.A.: RECENT ANALYTICAL DEVELOPMENTS IN THE PETROLEUM INDUSTRY, P.247, INSTITUTE OF PETROLEUM, LONDON (1974)
13. BARNETT D.B.; GREENFIELD A.A.; HOWLETT P.J.; HUDSON J.C.; SMITH R.M.: BRIT. MED. J., 2, 144 (1973)
14. BARON D.N.: ANN. CLIN. BIOCHEM., 7, 100 (1970)
15. BENDER C.F.: IN HADZI D.; ZUPAN J. (EDS.): COMPUTERS IN CHEMICAL RESEARCH AND EDUCATION, VOL. 2, P. 75, ELSEVIER, AMSTERDAM (1973)
16. BENDER C.F.; KOWALSKI B.R.: ANAL. CHEM., 45, 590 (1973)
17. BENDER C.F.; KOWALSKI B.R.: ANAL. CHEM., 46, 294 (1974)
18. BENDER C.F.; SHEPHERD H.D.; KOWALSKI B.R.: ANAL. CHEM., 45, 617 (1973)
19. BIRD J.R.; RUSSELL L.H.; SCOTT M.D.; AMBROSE W.R.: ANAL. CHEM., 50, 2082 (1978)
20. BOS M.: ANAL. CHIM. ACTA, 112, 65 (1979)
21. BOS M.; JASINK G.: ANAL. CHIM. ACTA, 103, 151 (1978)
22. BRENT D.A.; ROTH B.; BRUNNER T.R.: 8 TH INT. MASS SPECTROMETRY CONFERENCE, OSLO, NORWAY (1979)
23. BRIGGS P.L.; PRESS F.: NATURE, 268, 125 (1977)
24. BRISSEY G.F.; SPENCER R.B.; WILKINS C.L.: ANAL. CHEM., 51, 2295 (1979)
25. BRUGGER W.E.; JURS P.C.: J. AGRIC. FOOD CHEM., 25, 1158 (1977)
26. BRUNNER T.R.; WILKINS C.L.: 27TH PITTSBURGH CONFERENCE, ABSTRACT NO. 394 (1976)
27. BRUNNER T.R.; WILKINS C.L.; LAM T.F.; SOLTZBERG L.J.; KABERLINE S.L.: ANAL. CHEM., 48, 1146 (1976)
28. BRUNNER T.R.; WILKINS C.L.; WILLIAMS R.C.; MC-COMBIE P.J.: ANAL. CHEM., 47, 662 (1975)
29. BRUNNER T.R.; WILLIAMS R.C.; WILKINS C.L.; MC-COMBIE P.J.: ANAL. CHEM., 46, 1798 (1974)
30. BULBROOK R.D.; GREENWOOD F.C.; HAYWARD J.L.; SPICER C.C.: LANCET, 1960, 1154 (1960)

31. BURBANK F.: AM. J. MED., 46, 401 (1969)
32. BURGARD D.R.; PERONE S.P.: ANAL. CHEM., 50, 1366 (1978)
33. BURGARD D.R.; PERONE S.P.; WIEBERS J.L.: ANAL. CHEM., 49,
    1444 (1977)
34. BURGARD D.R.; PERONE S.P.; WIEBERS J.L.: BIOCHEMISTRY, 16,
    1051 (1977)
35. CAMMARATA A.; MENON G.K.: J. MED. CHEM., 19, 739 (1976)
36. CAREY R.N.; WOLD S.; WESTGARD J.O.: ANAL. CHEM., 47, 1824
    (1975)
37. CARLSON L.R.; BENDER C.F.; PRITCHARD R.H.: RESEARCH/
    DEVELOPMENT, FEBRUARY, 34 (1975)
38. CHAPMAN J.M.; COULSON A.H.; VIRGINIA A.C.; BORUN E.R.:
    J. CHRONIC. DIS., 23, 631 (1971)
39. CHAPMAN J.R.: COMPUTERS IN MASS SPECTROMETRY, P. 150,
    ACADEMIC PRESS, LONDON (1978)
40. CHOU J.T.; JURS P.C.: J. MED. CHEM., 22, 792 (1979)
41. CHRISTIE O.H.J.: ABSTRACTS OF THE 27TH IUPAC CONGRESS IN
    HELSINKI, P. 631 (1979)
42. CHU K.C.: ANAL. CHEM., 46, 1181 (1974)
43. CHU K.C.; FELDMANN R.J.; SHAPIRO M.B.; HAZARD G.F.JR.;
    GERAN R.I.: J. MED. CHEM., 18, 539 (1975)
44. CLARK H.A.; JURS P.C.: ANAL. CHEM., 47, 374 (1975)
45. CLARK H.A.; JURS P.C.: ANAL. CHEM., 51, 616 (1979)
46. CLERC J.T.: IN HADZI D.; ZUPAN J. (EDS.): COMPUTERS IN
    CHEMICAL RESEARCH AND EDUCATION, VOL. 2., P. 109,
    ELSEVIER, AMSTERDAM (1973)
47. CLERC J.T.: CHIMIA, 31, 353 (1977)
48. CLERC J.T.: IN LUDENA E.V.; SABELLI N.H.; WAHL A.C.(EDS.):
    COMPUTERS IN CHEMICAL EDUCATION AND RESEARCH, P. 393,
    MARCEL DEKKER, NEW YORK, 1977
49. CLERC J.T.; KUTTER M.; REINHARD M.; SCHWARZENBACH R.:
    J. CHROMATOG., 123, 271 (1976)
50. CLERC J.T.; MILNE G.W.A.; VARMUZA K.: ADVANCES IN MASS
    SPECTROMETRY, VOL. 7, 1111 (1978)
51. CLERC J.T.; NÄGELI P.; SEIBL J.: CHIMIA, 27, 639 (1973)
52. COMERFORD J.M.: DISS. ABSTR. INT., B35, 1 (1975)
53. COMERFORD J.M.; ANDERSON P.G.; SNYDER W.H.; KIMMEL H.S.:
    SPECTROCHIM. ACTA, 33A, 651 (1977)
54. COOMANS D.; BROECKAERT I.; JONCKHEER M.; LEPOUDRE R.;
    MASSART D.L.: ABSTRACTS OF THE 27TH IUPAC CONGRESS IN
    HELSINKI, P. 625 (1979)
55. COOMANS D.; JONCKHEER M.; MASSART D.L.; BROECKAERT I.;
    BLOCKX P.: ANAL. CHIM. ACTA, 103, 409 (1978)
56. COOMANS D.; MASSART D.L.; KAUFMAN L.: ANAL. CHIM. ACTA,
    112, 97 (1979)
57. COX L.A.JR.; BENDER C.F.: RECOG: A POLYALGORITHM FOR THE
    ANALYSIS OF GENERALIZED DATA SETS, LAWRENCE LIVERMORE
    LABORATORY, CALIFORNIA, REPORT NO. UCID-16915 (1975)
58. CRAMER R.D.III.; REDL G.; BERKOFF C.E.: J. MED. CHEM., 17,
    533 (1974)
59. CRAWFORD L.R.; MORRISON J.D.: ANAL. CHEM., 40, 1469 (1968)
60. CRAWFORD L.R.; MORRISON J.D.: ANAL. CHEM., 43, 1790 (1971)
61. CURRIE L.A.; FILLIBEN J.J.; DE-VOE J.R.: ANAL. CHEM., 44,
    497R (1972)
62. CURTIS M.: NTIS ACC. NO. ADA-043802 (1977)
63. CURTIS M.; STARKS S.A.: PITTSBURGH CONFERENCE, CLEVELAND,
    ABSTR. NO. 327 (1976)
64. DAMSGAARD E.; HEYDORN K.; LARSEN N.A.; PAKKENBERG H.;
    WOLD S.: ABSTRACTS OF THE 27TH IUPAC CONGRESS IN
    HELSINKI, P. 622 (1979)
65. DENISOV D.A.; DROZDOV-TICHOMIROV L.N.; GRIGORYEVA D.N.:
    J. THEOR. BIOL., 41, 431 (1973)

66. DE-BRUIN M.; KORTHOVEN P.J.M.; BAKELS C.C.; GROEN F.C.A.:
    ARCHAEOMETRY, 14, 55 (1972)
67. DE-CLERCQ H.; MASSART D.L.: J. CHROMATOGR., 115, 1 (1975)
68. DE-CLERCQ H.; MASSART D.L.; DRYON L.: J. PHARM. SCI., 66,
    1269 (1977)
69. DE-CLERCQ H.; VAN-OUDHEUSDEN D.; MASSART D.L.: ANALUSIS,
    3, 527 (1975)
70. DE-PALMA R.A.; PERONE S.P.: ANAL. CHEM., 51, 825 (1979)
71. DE-PALMA R.A.; PERONE S.P.: ANAL. CHEM., 51, 829 (1979)
72. DIERDORF D.S.; KOWALSKI B.R.: NTIS-REPORT, NO. AD-785863/
    2GA (1974)
73. DRACK H.: ANAL. CHEM., 50, 2147 (1978)
74. DRACK H.; KOSINA S.; GRASSERBAUER M.: FRESENIUS Z. ANAL.
    CHEM., 295, 30 (1979)
75. DROMEY R.G.: ANAL. CHEM., 48, 1464 (1976)
76. DROZDOV-TICHOMIROV L.N.: OPT. SPECTROSC., 27, 77 (1968)
77. DUEWER D.L.; FREISER H.: ANAL. CHEM., 49, 1940 (1977)
78. DUEWER D.L.; KOSKINEN J.R.; KOWALSKI B.R.: ARTHUR
    (PATTERN RECOGNITION PROGRAM), AVAILABLE FROM KOWALSKI B.R.,
    LAB. FOR CHEMOMETRICS, DEPT. OF CHEMISTRY, BG-10,
    UNIV. WASHINGTON, SEATTLE, WASH. 98195, USA (1975)
79. DUEWER D.L.; KOWALSKI B.R.: ANAL. CHEM., 47, 526 (1975)
80. DUEWER D.L.; KOWALSKI B.R.; SCHATZKI T.F.: ANAL. CHEM.,
    47, 1573 (1975)
81. DUNN W.J.; WOLD S.: J. MED. CHEM., 21, 1001 (1978)
82. DUNN W.J.; WOLD S.; MARTIN Y.C.: J. MED. CHEM., 21, 922
    (1978)
83. ESKES A.; DUPUIS P.F.; DIJKSTRA A.; DE-CLERCQ H.; MASSART
    D.L.: ANAL. CHEM., 47, 2168 (1975)
84. FELLOUS R.; LUFT R.; RABINE J.P.: J. CHROMATOGR., 166, 383
    (1978)
85. FELTY W.L.; JURS P.C.: ANAL. CHEM., 45, 885 (1973)
86. FRANZEN J.: CHROMATOGRAPHIA, 7, 518 (1974)
87. FRANZEN J.; HILLIG H.: ADVANCES IN MASS SPECTROMETRY, VOL.
    6, 991 (1974)
88. FREW N.M.: DISSERTATION, UNIV. WASHINGTON, DISS. ABSTR.,
    NO. 72-7354 (1972)
89. FREW N.M.; WANGEN L.E.; ISENHOUR T.L.: PATTERN
    RECOGNITION, 3, 281 (1971)
90. GAARENSTROOM P.D.; PERONE S.P.; MOYERS J.L.: ENVIRON. SCI.
    TECHNOL., 11, 795 (1977)
91. GIDDINGS J.C.: SEPARAT. SCI. TECHNOL., 13, 3 (1978)
92. GILLER S.A.; GLAZ A.B.; RASTRIGIN L.A.; ROSENBLIT A.B.:
    DOKL. AKAD. NAUK SSSR, 199, 851 (1971)
93. GOODRICH K.E.: MASTERS THESIS, UNIVERSITY OF WASHINGTON
    (1969)
94. GRAY N.A.B.: ANAL. CHEM., 48, 2265 (1976)
95. GRAY N.A.B.; GRONNEBERG T.O.: ANAL. CHEM., 47, 419 (1975)
96. GRIFFITHS A.G.: M. ENG. SCI. THESIS, JAMES COOK UNIVERSITY
    OF NORTH QUEENSLAND, TOWNSVILLE, AUSTRALIA (1977)
97. GUND P.: IN HAHN F.E.; KERSTEN H.; KERSTEN W.; SZYBALSKI
    W. (EDS.): PROGRESS IN MOLECULAR AND SUBCELLULAR BIOLOGY,
    VOL. 5, P. 117, SPRINGER, BERLIN (1977)
98. HAKEN J.K.; WAINWRIGHT M.S.; DO-PHUONG N.: J. CHROMATOGR.,
    117, 23 (1976)
99. HANSCH C.: IN CHAPMAN N.B., SHORTER J. (EDS.):
    CORRELATION ANALYSIS IN CHEMISTRY, P. 397, PLENUM PRESS,
    NEW YORK AND LONDON (1978)
100. HANSCH C.; UNGER S.H.; FORSYTHE A.B.: J. MED. CHEM., 16,
    1217 (1973)

101. HARPER A.M.; DUEWER D.L.; KOWALSKI B.R.; FASCHING J.L.:
     IN KOWALSKI B.R. (ED.): CHEMOMETRICS: THEORY AND
     APPLICATION, ACS SYMPOSIUM SERIES, NO. 52, P.14 (1977)
102. HARPER A.M.; KOWALSKI B.R.; BOYD J.; HESS G.: ABSTRACTS
     OF THE 27TH IUPAC CONGRESS IN HELSINKI, P. 623 (1979)
103. HARTMANN N.; HAWKES S.J.: J. CHROMATOGR. SCI., 8, 610
     (1970)
104. HEITE F.H.; DUPUIS P.F.; VANT-KLOOSTER H.A.; DIJKSTRA A.:
     ANAL. CHIM. ACTA, 103, 313 (1978)
105. HELLER S.R.: IN WIPKE W.T.; HELLER S.R.; FELDMANN R.J.;
     HYDE E. (EDS.): COMPUTER REPRESENTATION AND MANIPULATION
     OF CHEMICAL INFORMATION, P. 175, WILEY, NEW YORK, 1974
106. HELLER S.R.; CHANG C.L.; CHU K.C.: ANAL. CHEM., 46, 951
     (1974)
107. HILLER S.A.; GOLENDER V.E.; ROSENBLIT A.B.; RASTRIGIN L.A.;
     GLAZ A.B.: COMPUT. BIOCHEM. RES., 6, 411 (1973)
108. HILLIG H.: DISSERTATION, UNIV. DORTMUND, 1975
109. HODES L.; HAZARD G.F.; GERAN R.I.; RICHMAN S.: J. MED.
     CHEM., 20, 469 (1977)
110. HOPKE P.K.: J. ENVIRON. SCI. HEALTH A11, 367 (1976)
111. HOPKE P.K.; GLADNEY E.S.; GORDON G.E.; ZOLLER W.H.; JONES
     A.G.: ATMOS. ENVIRON., 10, 1015 (1976)
112. HOWERY D.G.; WEINER P.H.; BLINDER J.S.: J. CHROMATOG. SCI.,
     12, 366 (1974)
113. HUBER J.F.K.; REICH G.: ANAL. CHIM. ACTA, 122, 139 (1980)
114. IOFFE I.I.; FEDOROV V.S.; MUKHENBERG K.M.; FUKS I.S.:
     DOKL. AKAD. NAUK. SSSR, 189, 1290 (1969)
115. ISENHOUR T.L.; JURS P.C.: ANAL. CHEM., 43, 20A (1971)
116. ISENHOUR T.L.; JURS P.C.: IN HEPPLE P.(ED.): THE
     APPLICATIONS OF COMPUTER TECHNIQUES IN CHEMICAL RESEARCH,
     THE INSTITUTE OF PETROLEUM, LONDON (1972)
117. ISENHOUR T.L.; JURS P.C.: IN MATTSON J.S.; MARK H.B.JR.;
     MAC-DONALD H.C.JR. (EDS.): COMPUTER FUNDAMENTALS FOR
     CHEMISTS, (COMPUTERS IN CHEMISTRY AND INSTRUMENTATION,
     VOL. 1), MARCEL DEKKER, NEW YORK (1973)
118. ISENHOUR T.L.; KOWALSKI B.R.; JURS P.C.: CRITICAL REVIEWS
     IN ANALYTICAL CHEMISTRY, 4, 1 (1974)
119. JANISZEWSKI R.: DISSERTATION, AKADEMIE DER WISSENSCHAFTEN
     DER DDR, BERLIN (1973)
120. JURS P.C.: DISSERTATION, UNIV. WASHINGTON, DISS. ABSTR.,
     NO. 69-20241 (1969)
121. JURS P.C.: ANAL. CHEM., 42, 1633 (1970)
122. JURS P.C.: ANAL. CHEM., 43, 1812 (1971)
123. JURS P.C.: ANAL. CHEM., 43, 22 (1971)
124. JURS P.C.: APPL. SPECTROSC., 25, 483 (1971)
125. JURS P.C.: IN WIPKE W.T; HELLER S.R; FELDMANN R.J; HYDE E.
     (EDS.): COMPUTER REPRESENTATION AND MANIPULATION OF
     CHEMICAL INFORMATION, P. 265, WILEY, NEW YORK (1974)
126. JURS P.C.: INFORM. CHEM., 1975, 77 (1975)
127. JURS P.C.; CHOU J.T.; YUAN M.: J. MED. CHEM., 22, 476
     (1979)
128. JURS P.C.; ISENHOUR T.L.: CHEMICAL APPLICATIONS OF
     PATTERN RECOGNITION, WILEY, NEW YORK, (1975)
129. JURS P.C.; KOWALSKI B.R.; ISENHOUR T.L.: ANAL. CHEM., 41,
     21 (1969)
130. JURS P.C.; KOWALSKI B.R.; ISENHOUR T.L.; REILLEY C.N.:
     ANAL. CHEM., 41, 1949 (1969)
131. JURS P.C.; KOWALSKI B.R.; ISENHOUR T.L.; REILLEY C.N.:
     ANAL. CHEM., 41, 690 (1969)
132. JURS P.C.; KOWALSKI B.R.; ISENHOUR T.L.; REILLEY C.N.:
     ANAL. CHEM., 42, 1387 (1970)

133. JUSTICE J.B.JR.: DISSERTATION, UNIV. NORTH CAROLINA,
     CHAPEL HILL, DISS. ABSTR., NO. 74-26890 (1974)
134. JUSTICE J.B.JR.; ANDERSON D.N.; ISENHOUR T.L.; MARSHALL J.
     C.: ANAL. CHEM., 44, 2087 (1972)
135. JUSTICE J.B.JR.; ISENHOUR T.L.: ANAL. CHEM., 46, 223 (1974)
136. JUSTICE J.B.JR.; ISENHOUR T.L.: ANAL. CHEM., 47, 2286
     (1975)
137. KABERLINE S.L.; WILKINS C.L.: ANAL. CHIM. ACTA, 103, 417
     (1978)
138. KABERLINE S.L.; WILKINS C.L.: 29 TH PITTSBURGH CONFERENCE
     ANALYTICAL CHEMISTRY, CLEVELAND, ABSTR. NO. 660 (1978)
139. KAWAHARA F.K.; SANTNER J.F.; JULIAN E.C.: ANAL. CHEM., 46,
     266 (1974)
140. KAWAHARA F.K.; YANG Y.Y.: ANAL. CHEM., 48, 651 (1976)
141. KENT P.; GÄUMANN T.: HELV. CHIM. ACTA, 58, 787 (1975)
142. KHOTS M.S.: ZH. ANALIT. KHIM., 28, 797 (ENGLISH
     TRANSLATION: P. 710) (1973)
143. KHOTS M.S.; LERMAN A.G.; POLYAKOVA A.A.: ZH. ANALIT. KHIM.,
     36, 572 (ENGLISH TRANSLATION: P. 481) (1976)
144. KHOTS M.S.; YARMARKOV M.R.; POPOV A.A.: ZH. ANALIT. KHIM.,
     32, 566 (ENGLISH TRANSLATION: P. 453) (1977)
145. KIRSCHNER G.L.; KOWALSKI B.R.: IN ARIENS E.J. (ED.): DRUG
     DESIGN, VOL. 8, P.73, ACADEMIC, NEW YORK (1978)
146. KOSKINEN J.R.; KOWALSKI B.R.: J. CHEM. INF. COMPUT. SCI.,
     15, 119 (1975)
147. KOWALSKI B.R.: CHEM. TECH., 4, 300 (1974)
148. KOWALSKI B.R.: IN KLOPFENSTEIN C.E.; WILKINS C.L. (EDS.):
     COMPUTERS IN CHEMICAL AND BIOCHEMICAL RESEARCH, VOL. 2,
     ACADEMIC PRESS, NEW YORK (1974)
149. KOWALSKI B.R.: ANAL. CHEM., 47, 1152 A (1975)
150. KOWALSKI B.R.: J. CHEM. INF. COMPUT. SCI., 15, 201 (1975)
151. KOWALSKI B.R.: CHEMOMETRICS: THEORY AND APPLICATION, ACS
     SYMPOSIUM SERIES, NO. 52, AMERICAN CHEMICAL SOCIETY,
     WASHINGTON, D.C. (1977)
152. KOWALSKI B.R.; BENDER C.F.: ANAL. CHEM., 44, 1405 (1972)
153. KOWALSKI B.R.; BENDER C.F.: J. AM. CHEM. SOC., 94, 5632
     (1972)
154. KOWALSKI B.R.; BENDER C.F.: ANAL. CHEM., 45, 2234 (1973)
155. KOWALSKI B.R.; BENDER C.F.: J. AM. CHEM. SOC., 95, 686
     (1973)
156. KOWALSKI B.R.; BENDER C.F.: J. AM. CHEM. SOC., 96, 916
     (1974)
157. KOWALSKI B.R.; BENDER C.F.: NATURWISSENSCHAFTEN, 62, 10
     (1975)
158. KOWALSKI B.R.; BENDER C.F.: PATTERN RECOGNITION, 8, 1 (1976)
159. KOWALSKI B.R.; JURS P.C.; ISENHOUR T.L.; REILLEY C.N.:
     ANAL. CHEM., 41, 1945 (1969)
160. KOWALSKI B.R.; JURS P.C.; ISENHOUR T.L.; REILLEY C.N.:
     ANAL. CHEM., 41, 695 (1969)
161. KOWALSKI B.R.; REILLY C.A.: J. PHYS. CHEM., 75, 1402
     (1971)
162. KOWALSKI B.R.; SCHATZKI T.F.; STROSS F.H.: ANAL. CHEM.,
     44, 2176 (1972)
163. KRENMAYR P.; VARMUZA K.: ALLG. PRAKT. CHEM.(WIEN), 23,
     289 (1972)
164. KUO K.S.; JURS P.C.: J. AM. WATER WORKS ASSOC., 65, 623
     (1973)
165. LAM T.F.; WILKINS C.L.; BRUNNER T.R.; SOLTZBERG L.J.;
     KABERLINE S.L.: ANAL. CHEM., 48, 1768 (1976)
166. LEARY J.J.; JUSTICE J.B.JR.; TSUGE S.; LOWRY S.R.;
     ISENHOUR T.L.: J. CHROMATOGR. SCI., 11, 201 (1973)

167. LIDELL R.W.III.; JURS P.C.: APPL. SPECTROSC., 27, 371
     (1973)
168. LIDELL R.W.III.; JURS P.C.: ANAL. CHEM., 46, 2126 (1974)
169. LIN C.H.; CHEN H.F.: ANAL. CHEM., 49, 1357 (1977)
170. LOWRY S.R.: DISSERTATION, UNIV. NORTH CAROLINA, CHAPEL
     HILL (1975), DISS. ABSTR., NO. 76-20053 (1976)
171. LOWRY S.R.; ISENHOUR T.L.: J. CHEM. INF. COMPUT. SCI., 15,
     212 (1975)
172. LOWRY S.R.; ISENHOUR T.L.; JUSTICE J.B.JR.; MC-LAFFERTY F.W.;
     DAYRINGER H.E.; VENKATARAGHAVAN R.: ANAL. CHEM., 49,
     1720 (1977)
173. LOWRY S.R.; MARSHALL J.C.; ISENHOUR T.L.: COMPUT. CHEM.,
     1, 3 (1976)
174. LOWRY S.R.; RITTER G.L.; WOODRUFF H.B.; ISENHOUR T.L.:
     J. CHROMATOGR. SCI., 14, 126 (1976)
175. LOWRY S.R.; TSUGE S.; LEARY J.J.; ISENHOUR T.L.:
     J. CHROMATOGR. SCI., 12, 124 (1974)
176. LOWRY S.R.; WOODRUFF H.B.; RITTER G.L.; ISENHOUR T.L.:
     ANAL. CHEM., 47, 1126 (1975)
177. LYSYJ I.; NEWTON P.R.: ANAL. CHEM., 44, 2385 (1972)
178. LYTLE F.E.: ANAL. CHEM., 44, 1867 (1972)
179. MAC-DONALD J.C.: AMER. LAB., 9, 31 (1977)
180. MAC-DONALD J.C.: INTERNATIONAL LABORATORY, 1977, MARCH/
     APRIL, 37 (1977)
181. MAC-DONALD J.C.: INTERNATIONAL LABORATORY, 1978, MARCH/
     APRIL, 78 (1978)
182. MAC-DONALD J.C.; LANG G.E.JR.: INTERNATIONAL LABORATORY,
     1975, MARCH/APRIL, 9 (1975)
183. MAC-DONALD J.C.; LEVINE R.A.: PROCEEDINGS OF THE FOURTH
     NEW ENGLAND BIOENGINEERING CONFERENCE, PERGAMON, NEW
     YORK (1976)
184. MAHLE N.H.; ASHLEY J.W.: COMPUT. CHEM., 3, 19 (1979)
185. MARTIN Y.C.: QUANTITATIVE DRUG DESIGN: A CRITICAL
     INTRODUCTION (MEDICINAL RESEARCH, VOL. 8), MARCEL DEKKER,
     NEW YORK AND BASEL (1978)
186. MARTIN Y.C.; HOLLAND J.B.; JARBOE C.H.; PLOTNIKOFF N.: J.
     MED. CHEM., 17, 409 (1974)
187. MASSART D.L.; DE-CLERCQ H.: ANAL. CHEM., 46, 1988 (1974)
188. MASSART D.L.; DE-CLERCQ H.: ADVAN. CHROMATOGR., 16, 75
     (1978)
189. MASSART D.L.; DIJKSTRA A.; KAUFMAN L.: EVALUATION AND
     OPTIMIZATION OF LABORATORY METHODS AND ANALYTICAL
     PROCEDURES, ELSEVIER, AMSTERDAM (1978)
190. MASSART D.L.; KAUFMAN L.: ANAL. CHEM., 47, 1244A (1975)
191. MASSART D.L.; LENDERS P.; LAUWEREYS M.: J. CHROMATOGR. SCI.,
     12, 617 (1974)
192. MATHEWS R.J.: AUSTRALIAN J. CHEM., 26, 1955 (1973)
193. MATHEWS R.J.: INT. J. MASS SPECTROM. ION PHYS., 17, 217
     (1975)
194. MATHEWS R.J.: J. AM. CHEM. SOC., 97, 935 (1975)
195. MATTSON J.S.; MATTSON C.S.; SPENCER M.J.; SPENCER F.W.:
     ANAL. CHEM., 49, 500 (1977)
196. MC-CLOSKEY D.H.; HAWKES S.J.: J. CHROMATOGR. SCI., 13, 1
     (1975)
197. MC-GILL J.R.: DISSERTATION, UNIV. WASHINGTON, DISS. ABSTR.,
     NO. 77-26848 (1977)
198. MC-GILL J.R.; KOWALSKI B.R.: ANAL. CHEM., 49, 596 (1977)
199. MC-GILL J.R.; KOWALSKI B.R.: APPL. SPECTROSC., 31, 87 (1977)
200. MC-GILL J.R.; KOWALSKI B.R.: J. CHEM. INF. COMPUT. SCI.,
     18, 52 (1978)

201. MC-LAFFERTY F.W.: ANAL. CHEM., 49, 1441 (1977)
202. MC-LAFFERTY F.W.: PURE APPL. CHEM., 50, 197 (1978)
203. MC-LAFFERTY F.W.; VENKATARAGHAVAN R.: J. CHROMATOGR. SCI.,
     17, 24 (1979)
204. MECKE R.; NOAK K.: CHEM. BER., 93, 210 (1960)
205. MEISEL W.S.; JOLLEY M.; HELLER S.R.; MILNE G.W.A.: ANAL.
     CHIM. ACTA, 112, 407 (1979)
206. MELLON F.A.: IN JOHNSTONE R.A.W (ED.): MASS SPECTROMETRY,
     VOL. 3, P. 117, THE CHEMICAL SOCIETY, LONDON (1975)
207. MELLON F.A.: IN JOHNSTONE R.A.W (ED.): MASS SPECTROMETRY,
     VOL. 4, P. 59, THE CHEMICAL SOCIETY, LONDON (1977)
208. MELLON F.A.: IN JOHNSTONE R.A.W (ED.): MASS SPECTROMETRY,
     VOL. 5, P.100, THE CHEMICAL SOCIETY, LONDON (1979)
209. MENON G.K.; CAMMARATA A.: J. PHARM. SCI., 66, 304 (1977)
210. MERCOLINO T.J.; MAC-DONALD J.C.: 29 TH PITTSBURGH CONFERENCE
     ANALYTICAL CHEMISTRY, CLEVELAND, ABSTR. NO. 659 (1978)
211. MERRIT C.JR.; ROBERTSON D.H.; GRAHAM R.A.: ADVANCES IN
     MASS SPECTROMETRY, VOL. 7, 1002 (1978)
212. MEUZELAAR H.L.C.; KISTEMAKER P.G.; ESHUIS W.; BOERBOOM H.
     A.J.: ADVANCES IN MASS SPECTROMETRY, VOL. 7, 1452 (1978)
213. MIYASHITA Y.; ABE H.; SASAKI S.I.; YUTA K.: ANAL. CHEM.,
     50, 1580 (1978)
214. MORGAN D.R.: APPL. SPECTROSC., 31, 404 (1977)
215. MORGAN D.R.: APPL. SPECTROSC., 31, 415 (1977)
216. MURROW T.R.: M.S.THESIS, A.F. INST. TECH. WRIGHT-
     PATTERSON AIR FORCE BASE, GE/EE/71-20, OHIO (1971)
217. NÄGELI P.: DISSERTATION, ETH ZURICH, (1975)
218. NEUSTADTER H.E.; FORDYCE J.S.; KING R.B.: J. AIR POLL.
     CONTROL ASSOC., 26, 1079 (1976)
219. NORDEN B.; EDLUND U.; WOLD S.: ACTA CHEM. SCAND., B32,
     602 (1978)
220. OP-DE-BEEK J.; HOSTE J.: ANALYST, 99, 973 (1974)
221. PENCA M.; ZUPAN J.; HADZI D.: ANAL. CHIM. ACTA, 95, 3 (1977)
222. PERONE S.P.; PICHLER M.A.; GAARENSTROOM P.D.; MOYERS J.L.:
     INT. CONF. ON ENVIRONMENTAL SENSING AND ASSESSMENT, THE
     INSTITUTE OF ELECTRICAL AND ELECTRONICS ENGINEERS, INC.,
     USA (1976)
223. PERRIN C.L.: SCIENCE, 183, 551 (1974)
224. PESYNA G.M.: DISSERTATION, CORNELL UNIVERSITY (1975),
     DISS. ABSTR., NO. 76-12886 (1976)
225. PETERSEN E.M.: DISSERTATION, FORDHAM UNIV., NEW YORK,
     (1975), DISS. ABSTR., NO. 76-4137 (1976)
226. PETERSON J.T.: ATMOS. ENVIRON., 4, 501 (1970)
227. PETERSON J.T.: ATMOS. ENVIRON., 6, 433 (1972)
228. PICHLER M.A.; PERONE S.P.: ANAL. CHEM., 46, 1790 (1974)
229. PIETRANTONIO L.; JURS P.C.: PATTERN RECOGNITION, 4, 391
     (1972)
230. PIJPERS F.W.; VAN-GAAL H.L.M.; VAN-DER-LINDEN J.G.M.:
     ANAL. CHIM. ACTA., 112, 199 (1979)
231. POWERS J.J.; KEITH E.S.: J. FOOD SCI., 33, 207 (1968)
232. PRATT D.B.; MOORE C.B.; PARSONS M.L.; ANDERSON D.L.:
     RESEARCH/DEVELOPMENT, 1978, FEBR., P.52 (1978)
233. PREUSS D.R.; JURS P.C.: ANAL. CHEM., 46, 520 (1974)
234. RAMSOE K.; TYGSTRUP N.; WINKEL P.: SCAND. J. CLIN. LAB.
     INVEST., 26, 307 (1970)
235. RAZNIKOV V.V.; TALROZE V.L.: DOKL. AKAD. NAUK SSSR, 170,
     379 (1966)
236. REDL G.; CRAMER R.D.III.; BERKOFF C.E.: CHEM. SOC. REV.,
     3, 273 (1974)
237. REICH G.: DISSERTATION, UNIV. VIENNA (1978)

238. RESI H.G.; ALFSEN B.E.: EUR. SPECTROSC. NEWS, 1, NO. 3,
     P. 19 (1975)
239. RICHARDS J.A.; GRIFFITHS A.G.: ANAL. CHEM., 51, 1358 (1979)
240. RITTER G.L.; ISENHOUR T.L.: COMPUT. CHEM., 1, 145 (1977)
241. RITTER G.L.; ISENHOUR T.L.: COMPUT. CHEM., 1, 243 (1977)
242. RITTER G.L.; LOWRY S.R.; ISENHOUR T.L.; WILKINS C.L.:
     J. CHEM. INF. AND COMPUT. SCI., SUBMITTED (1978)
243. RITTER G.L.; LOWRY S.R.; WILKINS C.L.; ISENHOUR T.L.:
     ANAL. CHEM., 47, 1951 (1975)
244. RITTER G.L.; LOWRY S.R.; WOODRUFF H.B.; ISENHOUR T.L.:
     ANAL. CHEM., 48, 1027 (1976)
245. RITTER G.L.; WOODRUFF H.B.: ANAL. CHEM., 49, 2116 (1977)
246. RITTER G.L.; WOODRUFF H.B.; LOWRY S.R.; ISENHOUR T.L.:
     IEEE TRANS., IT-21, 665 (1975)
247. ROSSELL J.C.; FASCHING J.L.: 29 TH PITTSBURGH CONFERENCE
     ANALYTICAL CHEMISTRY, CLEVELAND, ABSTR. NO. 663 (1978)
248. ROTTER H.: DISSERTATION, TECHNISCHE UNIVERSITAET, VIENNA
     (1976)
249. ROTTER H.; VARMUZA K.: ORG. MASS SPECTROM., 10, 874
     (1975)
250. ROTTER H.; VARMUZA K.: ANAL. CHIM. ACTA, 95, 25 (1977)
251. ROTTER H.; VARMUZA K.: ANAL. CHIM. ACTA, 103, 61 (1978)
252. ROZETT R.W.; PETERSEN E.M.: ANAL. CHEM., 47, 1301 (1975)
253. ROZETT R.W.; PETERSEN E.M.: ANAL. CHEM., 47, 2377 (1975)
254. ROZETT R.W.; PETERSEN E.M.: ANAL. CHEM., 48, 817 (1976)
255. ROZETT R.W.; PETERSEN E.M.: AMER. LAB., 9, 107 (1977)
256. ROZETT R.W.; PETERSEN E.M.: ADVANCES IN MASS SPECTROMETRY,
     VOL. 7, 993 (1978)
257. SASAKI S.I.; ABE H.: MASS SPECTROSC. JAPAN, (SHITSURYO
     BUNSEKI), 20, 131 (1972)
258. SASAKI S.I.; ISHIDA Y.: JAPAN ANALYST (BUNSEKI KAGAKU),
     21, 1029 (1972)
259. SAXBERG B.E.H.; DUEWER D.L.; BOOKER J.L.; KOWALSKI B.R.:
     ANAL. CHIM. ACTA, 103, 201 (1978)
260. SCHECHTER J.; JURS P.C.: APPL. SPECTROSC., 27, 225 (1973)
261. SCHECHTER J.; JURS P.C.: APPL. SPECTROSC., 27, 30 (1973)
262. SCHIFFMAN S.S.: SCIENCE, 185, 112 (1974)
263. SCHRADER B.; STEIGNER E.: FRESENIUS Z. ANAL. CHEM., 254,
     177 (1971)
264. SHOENFELD P.S.; DE-VOE J.R.: ANAL. CHEM., 48, 403R (1976)
265. SIMON P.J.; GIESSEN B.C.; COPELAND T.R.: ANAL. CHEM., 49,
     2285 (1977)
266. SJÖSTROM M.; EDLUND U.: J. MAGN. RESON., 25, 285 (1977)
267. SJÖSTROM M.; KOWALSKI B.R.: ANAL. CHIM. ACTA, 112, 11 (1979)
268. SLAGLE J.R.; CHANG C.L.; HELLER S.R.: HEURISTICS
     LABORATORY, NATIONAL INST. OF HEALTH, BETHESDA, USA,
     PRIVAT COMMUNICATION (1974)
269. SMEYERS-VERBEKE J.; MASSART D.L.; COOMANS D.: J. ASSOC.
     OFF. ANAL. CHEM., 60, 1382 (1977)
270. SMITH D.H.: ANAL. CHEM., 44, 536 (1972)
271. SMITH D.H.; EGLINTON G.: NATURE, 235, 325 (1972)
272. SNEATH P.H.A.; SOKAL R.R.: NUMERICAL TAXONOMY, W.H.
     FREEMAN A. CO., SAN FRANCISCO (1973)
273. SOLBERG H.E.: SCAND. J. CLIN. LAB. INVEST., 35, 705 (1975)
274. SOLBERG H.E.; SKREDE S.; BLOMHOFF J.P.: SCAND. J. CLIN.
     LAB. INVEST., 35, 713 (1975)
275. SOLBERG H.E.; SKREDE S.; ELGJO K.; BLOMHOFF J.P.; GJONE E.:
     SCAND. J. CLIN. LAB. INVEST., 36, 81 (1976)
276. SOLTZBERG L.J.; WILKINS C.L.: J. AM. CHEM. SOC., 98, 4006
     (1976)

277. SOLTZBERG L.J.; WILKINS C.L.: 27TH PITTSBURGH CONFERENCE,
     CLEVELAND, ABSTRACT NO. 400, (1976)
278. SOLTZBERG L.J.; WILKINS C.L.: J. AM. CHEM. SOC., 99, 439
     (1977)
279. SOLTZBERG L.J.; WILKINS C.L.; KABERLINE S.L.; LAM T.F.;
     BRUNNER T.R.: J. AM. CHEM. SOC., 98, 7139 (1976)
280. SOLTZBERG L.J.; WILKINS C.L.; KABERLINE S.L.; LAM T.F.;
     BRUNNER T.R.: J. AM. CHEM. SOC., 98, 7144 (1976)
281. SOMMERAUER H.: DIPLOMARBEIT, EIDGEN. TECHNISCHE
     HOCHSCHULE ZÜRICH (1974)
282. STONHAM T.J.; ALEKSANDER I.: ELECTRON. LETT., 10, 301
     (1974)
283. STONHAM T.J.; ALEKSANDER I.; CAMP M.; PIKE W.T.; SHAW M.A.:
     ANAL. CHEM., 47, 1817 (1975)
284. STONHAM T.J.; ALEKSANDER I.; CAMP M.; SHAW M.A.; PIKE W.T.:
     ELECTRON. LETT., 9, 391 (1973)
285. STONHAM T.J.; SHAW M.A.: PATTERN RECOGNITION, 7, 235
     (1975)
286. STROUF O.; WOLD S.: ACTA CHEM. SCAND., A31, 391 (1977)
287. STUPER A.J.; BRUGGER W.E.; JURS P.C.: IN KOWALSKI B.R.
     (ED.): CHEMOMETRICS: THEORY AND APPLICATION, ACS
     SYMPOSIUM SERIES, NO. 52, P.165, (1977)
288. STUPER A.J.; BRUGGER W.E.; JURS P.C.: COMPUTER ASSISTED
     STUDIES OF CHEMICAL STRUCTURE AND BIOLOGICAL FUNCTION,
     WILEY, NEW YORK (1979)
289. STUPER A.J.; JURS P.C.: J. AM. CHEM. SOC., 97, 182 (1975)
290. STUPER A.J.; JURS P.C.: J. CHEM. INF. AND COMPUT. SCI.,
     16, 238 (1976)
291. STUPER A.J.; JURS P.C.: J. CHEM. INF. AND COMPUT. SCI.,
     16, 99 (1976)
292. STUPER A.J.; JURS P.C.: J. PHARM. SCI., 67, 745 (1978)
293. SWAIN T.: CHEMICAL PLANT TAXONOMY, ACADEMIC PRESS, LONDON
     AND NEW YORK (1963)
294. SYBRANDT L.B.; PERONE S.P.: ANAL. CHEM., 43, 382 (1971)
295. SYBRANDT L.B.; PERONE S.P.: ANAL. CHEM., 44, 2331 (1972)
296. THOMAS Q.V.; DE-PALMA R.A.; PERONE S.P.: ANAL. CHEM., 49,
     1376 (1977)
297. THOMAS Q.V.; PERONE S.P.: ANAL. CHEM., 49, 1369 (1977)
298. TING K.L.H.: SCIENCE, 183, 552 (1974)
299. TING K.L.H.; LEE R.C.T.; MILNE G.W.A.; SHAPIRO M.B.;
     GUARINO Å.M.: SCIENCE, 180, 417 (1973)
300. TUNNICLIFF D.D.; WADSWORTH P.A.: ANAL. CHEM., 45, 12
     (1973)
301. ULFVARSON U.; WOLD S.: SCAND. J. WORK. ENVIRON. A. HEALTH,
     3, 183 (1977)
302. UNGER S.H.: CANCER CHEMOTHERAP. REP., PART 2, 4, 45
     (1974)
303. VANDEGINSTE B.G.M.: ANAL. LETT., 10, 661 (1977)
304. VANDEGINSTE B.G.M.; VAN-LERSEL P.B.W.: PROC. ANALYT.
     DIV. CHEM. SOC., 15, 10 (1978)
305. VARMUZA K.: FRESENIUS Z. ANAL. CHEM., 268, 352 (1974)
306. VARMUZA K.: MONATSH. CHEM., 105, 1 (1974)
307. VARMUZA K.: MONATSH. CHEM., 107, 43 (1976)
308. VARMUZA K.: FRESENIUS Z. ANAL. CHEM., 286, 329 (1977)
309. VARMUZA K.: VESTNIK SLOVENSKEGA KEMIJSKEGA DRUSTVA
     (BULLETIN OF THE SLOVENIAN CHEMICAL SOCIETY, SUPPLEMENT),
     26, 61 (1979)
310. VARMUZA K.; KRENMAYR P.: FRESENIUS Z. ANAL. CHEM., 266,
     274 (1973)
311. VARMUZA K.; KRENMAYR P.: FRESENIUS Z. ANAL. CHEM., 271,
     22 (1974)

312. VARMUZA K.; ROTTER H.: MONATSH. CHEM., 107, 547 (1976)
313. VARMUZA K.; ROTTER H.: ADVANCES IN MASS SPECTROMETRY, VOL.
     7, 1099 (1978)
314. VARMUZA K.; ROTTER H.: ADVANCES IN MASS SPECTROMETRY, VOL.
     8, IN PRINT (1980)
315. VARMUZA K.; ROTTER H.; KRENMAYR P.: CHROMATOGRAPHIA, 7,
     522 (1974)
316. VERESS G.E.; PUNGOR E.: ABSTRACTS OF THE 27TH IUPAC
     CONGRESS IN HELSINKI, P. 116 (1979)
317. VINK J.; HEERMA W.; KAMERLING J.P.; VLIEGENTHART J.F.G.:
     ORG. MASS SPECTROM., 9, 536 (1974)
318. VOLKMANN P.: DISSERTATION, TECHNISCHE UNIVERSITAET,
     BERLIN, 1974
319. WANGEN L.E.: DISSERTATION, UNIV. WASHINGTON, DISS. ABSTR.,
     NO. 72-7428 (1971)
320. WANGEN L.E.; FREW N.M.; ISENHOUR T.L.: ANAL. CHEM., 43,
     845 (1971)
321. WANGEN L.E.; FREW N.M.; ISENHOUR T.L.; JURS P.C.: APPL.
     SPECTROSC., 25, 203 (1971)
322. WANGEN L.E.; ISENHOUR T.L.: ANAL. CHEM., 42, 737 (1970)
323. WARD S.D.: IN WILLIAMS D.H. (ED.): MASS SPECTROMETRY, VOL.
     1, P. 253, THE CHEMICAL SOCIETY, LONDON (1971)
324. WARD S.D.: IN WILLIAMS D.H. (ED.): MASS SPECTROMETRY, VOL.
     2, P. 264, THE CHEMICAL SOCIETY, LONDON (1973)
325. WEBB J.; KIRK K.A.; NIEDERMEIER W.; GRIGGS J.H.; TURNER M.E.;
     JAMES T.N.: J. MOL. CELLULAR CARDIOL., 6, 383 (1974)
326. WEGSCHEIDER W.; LEYDEN D.E.: ADVAN. X-RAY ANAL., 22, 357
     (1979)
327. WEISEL C.P.; FASCHING J.L.: ANAL. CHEM., 49, 2114 (1977)
328. WERNER M.; BROOKS S.H.; COHNEN G.: CLIN. CHEM., 18, 116
     (1972)
329. WHITE R.F.; LEWINSON T.M.: J. AM. STAT. ASSOC., 72, 271
     (1977)
330. WIEBERS J.L.; SHAPIRO J.A.; PERONE S.P.; BURGARD D.R.:
     PROCEEDINGS OF THE 23RD ANNUAL CONFERENCE ON MASS
     SPECTROMETRY AND ALLIED TOPICS, P. 515, HOUSTON (1975)
331. WILDING P.; KENDALL M.J.; HOLDER R.; GRIMES J.A.; FARR M.:
     CLIN. CHIM. ACTA, 64, 185 (1975)
332. WILKINS C.L.: J. CHEM. INF. COMPUT. SCI., 17, 242 (1977)
333. WILKINS C.L.; BRUNNER T.R.: ANAL. CHEM., 49, 2136 (1977)
334. WILKINS C.L.; ISENHOUR T.L.: ANAL. CHEM., 47, 1849 (1975)
335. WILKINS C.L.; JURS P.C.: IN GRIFFITHS P.R. (ED.):
     TRANSFORM TECHNIQUES IN CHEMISTRY, PLENUM PRESS, NEW
     YORK (1978)
336. WILKINS C.L.; WILLIAMS R.C.; BRUNNER T.R.; MC-COMBIE P.J.:
     J. AM. CHEM. SOC., 96, 4182 (1974)
337. WILLIAMS R.C.; SWANSON R.M.; WILKINS C.L.: ANAL. CHEM.,
     46, 1803 (1974)
338. WINKEL P.; AFIFI A.A.; CADY L.D.; WEIL M.H.; SHUBIN H.:
     J. CHRONIC. DIS., 24, 61 (1971)
339. WINKEL P.; RAMSOE K.; LYNGBYE J.; TYGSTRUP N.: CLIN. CHEM.,
     21, 71 (1975)
340. WOLD S.: J. CHROMATOG. SCI., 13, 525 (1975)
341. WOLD S.: 2ND FACSS-MEETING, INDIANAPOLIS, OCTOBER (1975)
342. WOLD S.: PATTERN RECOGNITION, 8, 127 (1976)
343. WOLD S.: IRIA-SEMINARS SERIES, LE CHESNAY, FRANCE, MARCH
     (1976)
344. WOLD S.: FIRST INTERNATIONAL SYMPOSIUM ON DATA ANALYSIS
     AND INFORMATICS, VERSAILLES, FRANCE, SEPTEMBER (1977)
345. WOLD S.: TECHNOMETRICS, 20, 397 (1978)

346. WOLD S.: REPORT OF THE INSTITUT OF CHEMISTRY, UNIVERSITY
     OF UMEA, SWEDEN, JULY (1978)
347. WOLD S.: ABSTRACTS OF THE 27TH IUPAC CONGRESS IN HELSINKI,
     P. 621 (1979)
348. WOLD S.; ANDERSSON K.: J. CHROMATOGR., 80, 43 (1973)
349. WOLD S.; SJÖSTROM M.: IN KOWALSKI B.R. (ED.):
     CHEMOMETRICS: THEORY AND APPLICATION, ACS SYMPOSIUM
     SERIES, NO. 52, P.243 (1977)
350. WOLD S.; SJÖSTROM M.: IN CHAPMAN N.B., SHORTER J. (EDS.):
     CORRELATION ANALYSIS IN CHEMISTRY, P. 1, PLENUM PRESS,
     NEW YORK AND LONDON (1978)
351. WOODRUFF H.B.; LOWRY S.R.; ISENHOUR T.L.: ANAL. CHEM., 46,
     2150 (1974)
352. WOODRUFF H.B.; LOWRY S.R.; ISENHOUR T.L.: APPL. SPECTROSC.,
     29, 226 (1975)
353. WOODRUFF H.B.; LOWRY S.R.; RITTER G.L.; ISENHOUR T.L.:
     ANAL. CHEM., 47, 2027 (1975)
354. WOODRUFF H.B.; MUNK M.E.: ANAL. CHIM. ACTA, 95, 13 (1977)
355. WOODRUFF H.B.; RITTER G.L.; LOWRY S.R.; ISENHOUR T.L.:
     TECHNOMETRICS, 17, 455 (1975)
356. WOODRUFF H.B.; RITTER G.L.; LOWRY S.R.; ISENHOUR T.L.:
     APPL. SPECTROSC., 30, 213 (1976)
357. WOODRUFF H.B.; SNELLING C.R.JR.; SHELLEY C.A.; MUNK M.E.:
     ANAL. CHEM., 49, 2075 (1977)
358. WOODWARD W.S.; ISENHOUR T.L.: ANAL. CHEM., 46, 422 (1974)
359. ZANDER G.S.; JURS P.C.: ANAL. CHEM., 47, 1562 (1975)
360. ZANDER G.S.; STUPER A.J.; JURS P.C.: ANAL. CHEM., 47,
     1085 (1975)
361. ZIEGLER E.: COMPUTER IN DER INSTRUMENTELLEN ANALYTIK, P.
     187, AKADEMISCHE VERLAGSGESELLSCHAFT, FRANKFURT/MAIN (1973)
362. ZIEMER J.N.; PERONE S.P.; CAPRIOLI R.M.; SEIFERT W.E.:
     ANAL. CHEM., 51, 1732 (1979)
363. ZUPAN J.: ANAL. CHIM. ACTA, 103, 273 (1978)
364. ZUPAN J.; KOBE J.; HADZI D.: ANAL. CHEM., 47, 1702 (1975)

21.2. General Pattern Recognition

365. ABDALI S.K.: PATTERN RECOGNITION, 3, 3 (1971)
366. ALEKSANDER I.: IN BATCHELOR B.G. (ED.): PATTERN
     RECOGNITION - IDEAS IN PRACTICE, P. 43, PLENUM PRESS,
     NEW YORK (1978)
367. ALEKSANDER I.; ALBROW R.C.: COMPUT. J., 11, 65 (1968)
368. ANDREWS H.C.: INTRODUCTION TO THE MATHEMATICAL TECHNIQUES
     OF PATTERN RECOGNITION, WILEY, NEW YORK (1972)
369. ARKADJEW A.G.; BRAWERMANN E.M.: ZEICHENERKENNUNG UND
     MASCHINELLES LERNEN, OLDENBOURG VERLAG, MÜNCHEN UND WIEN,
     (1966)
370. BATCHELOR B.G.: PRACTICAL APPROACH TO PATTERN
     CLASSIFICATION, PLENUM PUBL. CORP., LONDON (1974)
371. BATCHELOR B.G.: (ED.): PATTERN RECOGNITION - IDEAS IN
     PRACTICE, PLENUM PRESS, NEW YORK (1978)
372. BLEDSOE W.V.; BROWNING I.: PROC. EASTERN JOINT COMPUTER
     CONFERENCE, P. 225 (1959)
373. CHERNOFF H.: J. AM. STATISTIC. ASSOC., 68, 361 (1973)
374. COVER T.M.; HART P.E.: IEEE TRANS. IT-13, 21 (1967)
375. DUBES R.; JAIN A.K.: PATTERN RECOGNITION, 11, 235 (1979)
376. DUDA R.O.; HART P.E.: PATTERN CLASSIFICATION AND SCENE
     ANALYSIS, WILEY, NEW YORK (1973)

377. EVERITT B.S.: GRAPHICAL TECHNIQUES FOR MULTIVARIATE DATA, HEINEMANN EDUCATIONAL BOOKS, LONDON (1978)
378. FEIGENBAUM E.; FELDMAN J.: (EDS.): COMPUTERS AND THOUGHT, MC GRAW HILL, NEW YORK (1963)
379. FIX E.; HODGES J.L.JR.: PROJECT NO. 21-49-004, REPORT NO. 4, PREPARED AT THE UNIVERSITY OF CALIFORNIA UNDER CONTRACT NO. AF 41(128)-31. USAF SCHOOL OF AVIATION MEDICINE, RANDOLPH FIELD, TEXAS (1951)
380. FUKUNAGA K.: INTRODUCTION TO STATISTICAL PATTERN RECOGNITION, ACADEMIC PRESS, NEW YORK (1972)
381. FUKUNAGA K.; OLSEN D.R.: IEEE TRANS., C-20, 917 (1971)
382. GATES G.W.: IEEE TRANS., IT-18, 431 (1972)
383. HART P.E.: IEEE TRANS., IT-14, 515 (1968)
384. HIGHLEYMAN W.H.: BELL SYST. TECH. J., 41, 723 (1962)
385. KANAL L.: IEEE TRANS. IT-20, 697 (1974)
386. KRUSKAL J.B.JR.: PROC. AMER. MATH. SOC., 7, 48 (1956)
387. LEVINE M.D.: PROC. IEEE, 57, 1391 (1969)
388. MC-CABE G.P.JR.: TECHNOMETRICS, 17, 103 (1975)
389. MEISEL W.S.: COMPUTER-ORIENTED APPROACHES TO PATTERN RECOGNITION, ACADEMIC PRESS, NEW YORK (1972)
390. MENDEL J.M.; FU K.S.: (EDS.): ADAPTIVE, LEARNING AND PATTERN RECOGNITION SYSTEMS, ACADEMIC PRESS, NEW YORK (1970)
391. MEYER-BRÖTZ G.; SCHÜRMANN J.: METHODEN DER AUTOMATISCHEN ZEICHENERKENNUNG, OLDENBOURG VERLAG, MÜNCHEN (1970)
392. MINSKY M.: PROC. IRE, 49, 8 (1961)
393. MINSKY M.; PAPERT S.: PERCEPTRONS - AN INTRODUCTION TO COMPUTATIONAL GEOMETRY, MIT PRESS, CAMBRIDGE, MASS. (1969)
394. NAGY G.: PROC. IRE, 56, 836 (1968)
395. NIEMANN H.: METHODEN DER MUSTERERKENNUNG, AKADEMISCHE VERLAGSGESELLSCHAFT, FRANKFURT/MAIN (1974)
396. NILSSON N.J.: LEARNING MACHINES, MC GRAW HILL, NEW YORK (1965)
397. PATRICK E.A.: FUNDAMENTALS OF PATTERN RECOGNITION, PRENTICE HALL, ENGLEWOOD CLIFFS, N.J. (1972)
398. ROGERS D.J.; TANIMOTO T.T.: SCIENCE, 132, 1115 (1960)
399. ROSENBLATT F.: PROC. IRE, 48, 301 (1960)
400. ROSEN C.A.: SCIENCE, 156, 38 (1967)
401. SAMMON J.W.JR.: IEEE TRANS., C-18, 401 (1969)
402. STEINHAGEN H.E.; FUCHS S.: OBJEKTERKENNUNG - EINFUEHRUNG IN DIE MATHEMATISCHEN METHODEN DER ZEICHENERKENNUNG, VEB-VERLAG TECHNIK, BERLIN (DDR), (1976)
403. TOU J.T.; GONZALES R.C.: PATTERN RECOGNITION PRINCIPLES, ADDISON-WESLEY PUBL. COMP. INC., READING (MASS., USA), (1974)
404. VERHAGEN C.J.D.M.: PATTERN RECOGNITION, 7, 109 (1975)
405. WAGNER T.J.: IEEE TRANS., IT-19, 696 (1973)
406. WILSON D.L.: IEEE TRANS., SMC-2, 408 (1972)
407. YOUNG T.Y.; CALVERT T.W.: CLASSIFICATION, ESTIMATION AND PATTERN RECOGNITION, ELSEVIER, NEW YORK (1974)
408. ZAHN C.T.: IEEE TRANS., C-20, 68 (1971)

21.3. Other Literature

409. BENZ W.: MASSENSPEKTROMETRIE ORGANISCHER VERBINDUNGEN, AKADEMISCHE VERLAGSGESELLSCHAFT, FRANKFURT/MAIN (1969)
410. BRILLOUIN L.: SCIENCE AND INFORMATION THEORY, ACADEMIC PRESS, NEW YORK (1956)
411. BRUGGER W.E.; STUPER A.J.; JURS P.C.: J. CHEM. INF. COMPUT. SCI., 16, 105 (1976)
412. CLEIJ P.; DIJKSTRA A.: FRESENIUS Z. ANAL. CHEM., 298, 97 (1979)
413. CLERC J.T.: PRIVAT COMMUNICATION (1975)
414. DEMING S.N.; MORGAN S.L.: ANAL. CHEM., 45, 278A (1973)
415. ERNST R.R.: REV. SCI. INSTR., 39, 998 (1968)
416. FISHER R.A.: ANN. EUGENICS, 7, 179 (1936)
417. GOLDMAN S.: INFORMATION THEORY, PRENTICE HALL, NEW YORK (1953)
418. GROTCH S.L.: ANAL. CHEM., 42, 1214 (1970)
419. HARTLEY R.V.L.: BELL. SYST. TECHN. J., 7, 535 (1928)
420. JAGLOM A.M.; JAGLOM I.M.: WAHRSCHEINLICHKEIT UND INFORMATION, VEB DEUTSCHER VERLAG DER WISSENSCHAFTEN, BERLIN (DDR) (1965)
421. KAISER H.: ANAL. CHEM., 42 [2], 24A (1970)
422. KWOK K.S.; VENKATARAGHAVAN R.; MC-LAFFERTY F.W.: J. AM. CHEM. SOC., 95, 4185 (1973)
423. MC-REYNOLD W.O.: J. CHROMATOGR. SCI., 8, 685 (1970)
424. MORGAN S.L.; DEMING S.N.: ANAL. CHEM., 46, 1170 (1974)
425. NELDER J.A.; MEAD R.: COMPUT. J., 7, 308 (1965)
426. OLSSON D.M.; NELSON L.S.: TECHNOMETRICS, 17, 45 (1975)
427. PETERS J.: EINFUEHRUNG IN DIE ALLGEMEINE INFORMATIONSTHEORIE, SPRINGER VERLAG, BERLIN (1967)
428. RAO C.R.: LINEAR STATISTICAL INFERENCE AND ITS APPLICATIONS, WILEY, NEW YORK (1965)
429. RAO C.R.: LINEARE STATISTISCHE METHODEN UND IHRE ANWENDUNGEN, AKADEMIE VERLAG, BERLIN (1973)
430. RENYI A.: WAHRSCHEINLICHKEITSRECHNUNG, VEB DEUTSCHER VERLAG DER WISSENSCHAFTEN, BERLIN (DDR) (1962)
431. REVENSTORF D.: LEHRBUCH DER FAKTORENANALYSE, VERLAG W. KOHLHAMMER, STUTTGART (1976)
432. ROUTH M.W.; SWARTZ P.A.; DENTON M.B.: ANAL. CHEM., 49, 1422 (1977)
433. SHANNON C.E.: BELL SYST. TECHN. J., 27, 379 (1948)
434. SHANNON C.E.: BELL SYST. TECHN. J., 27, 623 (1948)
435. SPENDLEY W.; HEXT G.R.; HIMSWORTH F.R.: TECHNOMETRICS, 4, 441 (1962)

21.4. List of Authors

Author	Reference Numbers
1. ABDALI S.K.	365
2. ABE H.	1 2 213 257
3. ADAMS R.A.	3
4. ADAMSON G.W.	4 5 6
5. AFIFI A.A.	338
6. ALBANO C.	7
7. ALBROW R.C.	367
8. ALEKSANDER I.	282 283 284 366 367
9. ALFSEN B.E.	238
10. AMBROSE W.R.	19
11. ANDERSON D.L.	232
12. ANDERSON D.N.	8 134
13. ANDERSON P.G.	53
14. ANDERSSON K.	348
15. ANDREEVA L.N.	9
16. ANDREWS H.C.	368
17. ARKADJEW A.G.	369
18. ARTEMOVA F.SH.	9
19. ASHLEY J.W.	184
20. BAER C.D.	10
21. BAILEY A.	11
22. BAKELS C.C.	66
23. BAKER A.G.	12
24. BARNETT D.B.	13
25. BARON D.N.	14
26. BATCHELOR B.G.	370 371
27. BENDER C.F.	15 16 17 18 37 57 152 153 154 155
	156 157 158
28. BENZ W.	409
29. BERKOFF C.E.	58 236
30. BIRD J.R.	19
31. BLEDSOE W.V.	372
32. BLINDER J.S.	112
33. BLOCKX P.	55
34. BLOMHOFF J.P.	274 275
35. BOERBOOM H.A.J.	212
36. BOOKER J.L.	259
37. BORUN E.R.	38
38. BOS M.	20 21
39. BOYD J.	102
40. BRAWERMANN E.M.	369
41. BRENT D.A.	22
42. BRIGGS P.L.	23
43. BRILLOUIN L.	410
44. BRISSEY G.F.	24
45. BROECKAERT I.	54 55
46. BROOKS S.H.	328
47. BROWN C.W.	10
48. BROWNING I.	372
49. BRUGGER W.E.	25 287 288 411
50. BRUNNER T.R.	22 26 27 28 29 165 279 280 333 336
51. BULBROOK R.D.	30
52. BURBANK F.	31
53. BURGARD D.R.	32 33 34 330
54. BUSH J.A.	4 5 6

461. YUAN M.   127
462. YUTA K.   213
463. ZAHN C.T.   408
464. ZANDER G.S.   359 360
465. ZIEGLER E.   361
466. ZIEMER J.N.   362
467. ZOLLER W.H.   111
468. ZUPAN J.   221 363 364

Year	Number of References
1928	1
1936	1
1948	2
1951	1
1953	1
1956	2
1959	1
1960	4
1961	1
1962	3
1963	2
1965	4
1966	2
1967	3
1968	7
1969	13
1970	12
1971	20
1972	25
1973	36
1974	55
1975	62
1976	42
1977	57
1978	46
1979	30
1980	2

22. Subject Index

THEORETICA CHIMICA ACTA

an International Journal
of Theoretical Chemistry

ISSN 0040-5744   Title No. 214

Edenda curat: Hermann Hartmann, Mainz

Adiuvantibus: C. J. Ballhausen, København; R. D. Brown, Clayton; K. Fukui, Kyoto; R. Gleiter, Heidelberg; E. A. Halevi, Haifa; G. G. Hall, Nottingham; E. Heilbronner, Basel; J. Jortner, Tel-Aviv; M. Kotani, Tokyo; J. Koutecký, Berlin; A. Neckel, Wien; E. E. Nikitin, Moskwa; R. G. Pearson, Santa Barbara; B. Pullmann, Paris; B. Rånby, Stockholm; K. Ruedenberg, Ames; C. Sandorfy, Montreal; M. Simonetta, Milano; O. Sinanoğlu, New Haven; R. Zahradník, Praha

Today, theory and experiment are inseparably bound. Every chemical experiment is preceded by reflection and careful consideration, and the results are interpreted according to chemical theories and perceptions.

The editors of **Theoretica Chimica Acta** therefore wish to emphasize the wide-ranging program reflected in the policy of their journal:

"**Theoretica Chimica Acta** accepts manuscripts in which the relationships between individual chemical and physical phenomena are investigated. In addition, experimental research that presents new theoretical viewpoints is desired."

Theoretica Chimica Acta offers experimental chemists increased space for the publication of discussion of the goals of their work, the significance of their findings, and the concepts on which their experimental work is based. Such discussions contribute significantly to mutual understanding between theoreticians and experimentalists and stimulate both new reflections and further experiments.

Springer
International

Subscription Information and/or sample copies upon request. Please send your order or request to your bookseller or directly to:
Springer-Verlag, Journal Promotion Department, P. O. Box 105280, D-6900 Heidelberg, FRG

A. F. Williams

A Theoretical Approach to Inorganic Chemistry

1979. 144 figures, 17 tables. XII, 316 pages
ISBN 3-540-09073-8

This book is intended to outline the application of simple quantum mechanics to the study of inorganic chemistry, and to show its potential for systematizing and understanding the structure, physical properties, and reactivities of inorganic compounds. The considerable development of inorganic chemistry in recent years necessitates the establishment of a theoretical framework if the student is to acquire sound knowledge of the subject. An effort has been made to cover a wide range of subjects, and to encourage the reader to think of further extensions of the theories discussed. The importance of the critical application of theory is emphasized, and, although the book is concerned chiefly with molecular orbital theory, other approaches are discussed. The book is intended for students in the latter half of their undergraduate studies.

Springer-Verlag
Berlin
Heidelberg
New York

Contents: Quantum Mechanics and Atomic Theory. – Simple Molecular Orbital Theory. – Structural Applications of Molecular Orbital Theory. – Electronic Spectra and Magnetic Properties of Inorganic Compounds. – Alternative Methods and Concepts. – Mechanism and Reactivity. – Descriptive Chemistry. – Physical and Spectroscopic Methods. – Appendices. – Subject Index.

Lecture Notes in Chemistry